BOY SCOUTS OF AMERICA
MERIT BADGE SERIES

OCEANOGRAPHY

"Enhancing our youths' competitive edge through merit badges"

Requirements

1. Name four branches of oceanography. Describe at least five reasons why it is important for people to learn about the oceans.
2. Define salinity, temperature, and density, and describe how these important properties of seawater are measured by the physical oceanographer. Discuss the circulation and currents of the ocean. Describe the effects of the oceans on weather and climate.
3. Describe the characteristics of ocean waves. Point out the differences among the storm surge, tsunami, tidal wave, and tidal bore. Explain the difference between sea, swell, and surf. Explain how breakers are formed.
4. Draw a cross-section of underwater topography. Show what is meant by:
   a. Continental shelf
   b. Continental slope
   c. Abyssal plain

   Name and put on your drawing the following: seamount, guyot, rift valley, canyon, trench, and oceanic ridge. Compare the depths in the oceans with the heights of mountains on land.
5. List the main salts, gases, and nutrients in seawater. Describe some important properties of water. Tell how the animals and plants of the ocean affect the chemical composition of seawater. Explain how differences in evaporation and precipitation affect the salt content of the oceans.

35924
ISBN 978-0-8395-3306-1
©2012 Boy Scouts of America
2015 Printing

6. Describe some of the biologically important properties of seawater. Define benthos, nekton, and plankton. Name some of the plants and animals that make up each of these groups. Describe the place and importance of phytoplankton in the oceanic food chain.

7. Do ONE of the following:

   a. Make a plankton net. Tow the net by a dock, wade with it, hold it in a current, or tow it from a rowboat.* Do this for about 20 minutes. Save the sample. Examine it under a microscope or high-power glass. Identify the three most common types of plankton in the sample.

   b. Make a series of models (clay or plaster and wood) of a volcanic island. Show the growth of an atoll from a fringing reef through a barrier reef. Describe the Darwinian theory of coral reef formation.

   c. Measure the water temperature at the surface, midwater, and bottom of a body of water four times daily for five consecutive days.* You may measure depth with a rock tied to a line. Make a Secchi disk to measure turbidity (how much suspended sedimentation is in the water). Measure the air temperature. Note the cloud cover and roughness of the water. Show your findings (air and water temperature, turbidity) on a graph. Tell how the water temperature changes with air temperature.

   d. Make a model showing the inshore sediment movement by littoral currents, tidal movement, and wave action. Include such formations as high and low waterlines, low-tide terrace, berm, and coastal cliffs. Show how offshore bars are built up and torn down.

   e. Make a wave generator. Show reflection and refraction of waves. Show how groins, jetties, and breakwaters affect these patterns.

   f. Track and monitor satellite images available on the Internet for a specific location for three weeks. Describe what you have learned to your counselor.

*May be done in lakes or streams.

8. Do ONE of the following:
   a. Write a 500-word report on a book about oceanography approved by your counselor.
   b. Visit one of the following:
      (1) Oceanographic research ship
      (2) Oceanographic institute, marine laboratory, or marine aquarium

      Write a 500-word report about your visit.
   c. Explain to your troop in a five-minute prepared speech "Why Oceanography Is Important" or describe "Career Opportunities in Oceanography." (Before making your speech, show your speech outline to your counselor for approval.)
9. Describe four methods that marine scientists use to investigate the ocean, underlying geology, and organisms living in the water.

Contents

Our Blue Planet

In some way, the oceans touch every part of Earth. The oceans cover more than 70 percent of our planet and are *the* dominant feature of Earth. Wherever you live, the oceans influence the weather, the soil, the air, and the geography of your community. *Apollo* astronauts hurtling toward the moon looked back at a blue sphere speckled with clouds. To study the oceans is to study Earth itself.

Since the world's oceans are all connected, you could think of them as one great ocean. But people have given the various oceans names and have even debated their number. They are, from largest to smallest, the Pacific, the Atlantic, the Indian, the Southern, and the Arctic Oceans.

What Is Oceanography?

Oceanography covers all aspects of ocean study and exploration.

- **Geological** oceanography focuses on the topographic features and physical makeup of the ocean floor.
- **Physical** oceanography deals with the motions of seawater, such as waves, tides, and currents.
- **Chemical** oceanography concerns the distribution of chemical compounds and chemical reactions in the ocean and on the seafloor.
- **Meteorological** oceanography pertains to the study of the ocean's interaction with the atmosphere and its effect on weather and climate.
- **Biological** oceanography concentrates on plant and animal life in the sea.

Studying the oceans tells us much about the land, rivers, lakes, and the air—our entire planet. This may help us to find new sources or supplies of food, freshwater, minerals, and energy, and a new understanding of weather and climatic patterns.

The International Hydrographic Organization lists the five oceans mentioned here, but some people (and maps) still use different names. Some people call the waters at the northernmost reaches of Earth the Arctic Sea and consider it part of the Atlantic Ocean. Some consider the Southern Ocean—which surrounds Antarctica—to be the southern reaches of the Atlantic, Pacific, and Indian Oceans, while others prefer to call it the Antarctic Ocean.

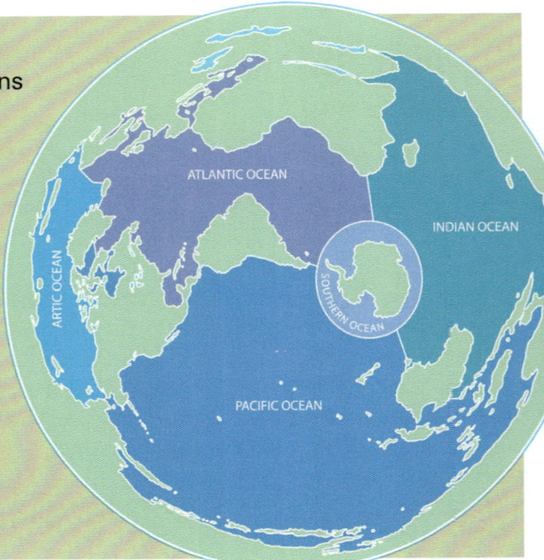

Connected to each ocean are seas, bays, and gulfs that you would consider part of that ocean. For example, the Red Sea, northeast of Africa, is part of the Indian Ocean, just as the Caribbean and Mediterranean Seas are part of the Atlantic. The Southern Ocean is the locale of some of the roughest seas on Earth.

The real beginning of deep-sea research came with the HMS *Challenger* expedition. In 1872, the *Challenger* left England with a crew of five scientists, 23 officers, and 243 sailors to explore the deep seas. For three and a half years, the vessel crossed the Atlantic, Pacific, and Antarctic (Southern) Oceans.

The crew's scientific research included measuring water temperatures at great depths and collecting sediments, water samples, and thousands of forms of marine life never before seen. The expedition brought new knowledge of ocean temperatures, ocean currents, and the depths and contours of the ocean basins. Scientists took 23 years to compile the results of the voyage in a 50-volume, 29,500-page report that is still used today.

Significance of the Challenger Expedition

The historical 1872 Challenger expedition led to:

- The first systematic plot of ocean temperatures and currents
- Development of the first maps of the deep-sea bottom deposits and water depths
- Discovery of the Mid-Atlantic Ridge
- Recording of water depth of 27,060 feet (8,248 meters) at the Challenger Deep in the Mariana Trench, the deepest known point in the world's oceans*
- Discovery of 715 new genera and 4,717 new species of ocean life forms, including phenomenal organisms living at great depths (proving that life exists even in extreme environments)

*Revised estimates of water depth there have ranged from about 35,800 to 36,200 feet (10,912 to 11,034 meters).

Throughout human history people have sought to understand the sea. The ancient Greeks spoke with awe of the god Poseidon, who they believed ruled the waves and the depths beneath. Polynesian sailors crossed thousands of miles on open-air crafts to settle a myriad of islands of the Pacific. European explorers plied the mysterious Atlantic westward in search of the herbs and spices of the Orient and eventually discovered what is now known as North and South America.

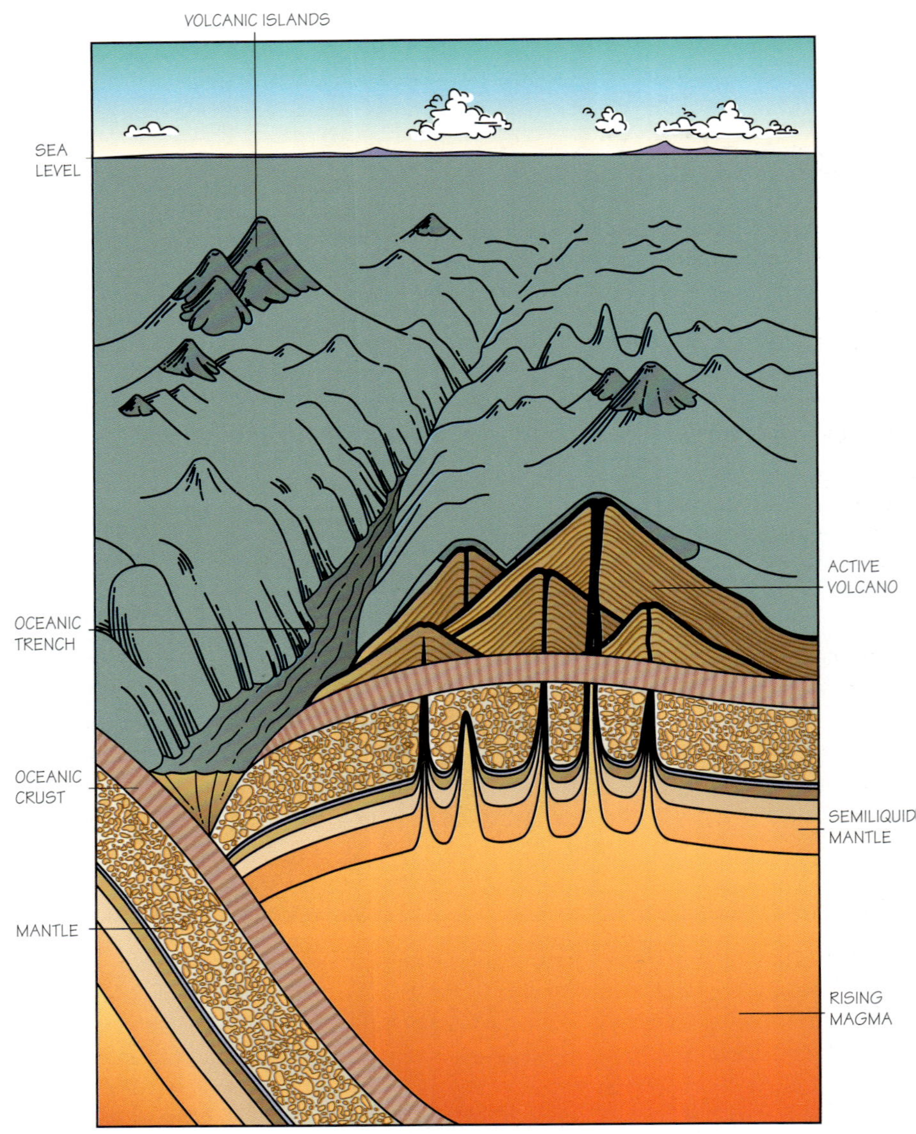

Underwater physical features of the ocean

Our Solid Earth

Many scientists believe that billions of years ago Earth had just one or two supercontinents, which slowly broke apart and moved away from one another to form the continents we know today. These same scientists believe that the outermost shell of Earth (the *lithosphere*) is formed of rigid plates that, over time, slowly move across Earth's surface.

Plate Tectonics

Much geological proof supports the idea of *continental drift*—that today's continents slowly move atop massive plates. The plates, perhaps 50 miles thick and up to thousands of miles across, float on a bed of partly molten rock. These plates extend under the ocean as well as under the continents.

"Plate tectonics" is a geological theory that holds that these plates slowly collide to form the world's mountain chains. Scientists have identified at least six major plates as well as some smaller ones. Plates pushing together or sliding under each other form *mid-ocean ridges,* or underwater mountain ranges, and *oceanic trenches.*

Oceanic Fracture Zones

Around the world, earthquakes most commonly happen along plate boundaries. Deep earthquakes occur where plates slide under each other. Shallow earthquakes occur along ocean ridges, or where plates slide by each other without colliding. *Oceanic fracture zones* are long, straight ridges and troughs that cut across the ocean ridges. Frequent underwater earthquakes cause new oceanic crust to form at these zones.

Rifts and Seafloor Spreading

Some narrow seas on Earth may be widening into new oceans. For example, the Red Sea between mainland Egypt and the Sinai Peninsula fills a widening rift in the continental crust. Many scientists believe that rift valleys indicate where continents are drifting apart prior to separating. This process of a crack widening under the sea is called *seafloor spreading.*

Magma (molten rock) from deep within Earth may emerge from an ocean rift and form a new seafloor basin. If this occurs and seafloor spreading continues, a new ocean may form. Evidence of rifts and sediments as well as volcanic formations that rim the coasts from New England to Iceland and Africa point to the formation of the Atlantic Ocean.

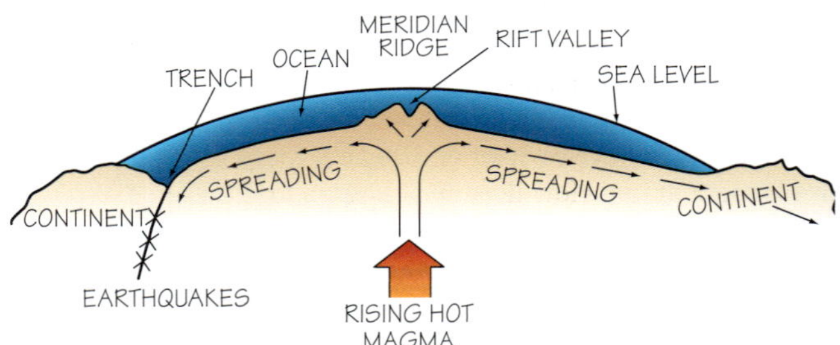

Seafloor spreading

The world's oceans are in different stages of widening and narrowing:

- The Atlantic Ocean is spreading on both sides.
- The Indian Ocean is widening on the west side but narrowing on the east.
- The Pacific Ocean is narrowing on both sides and probably will disappear in the future as Asia collides with the Americas.

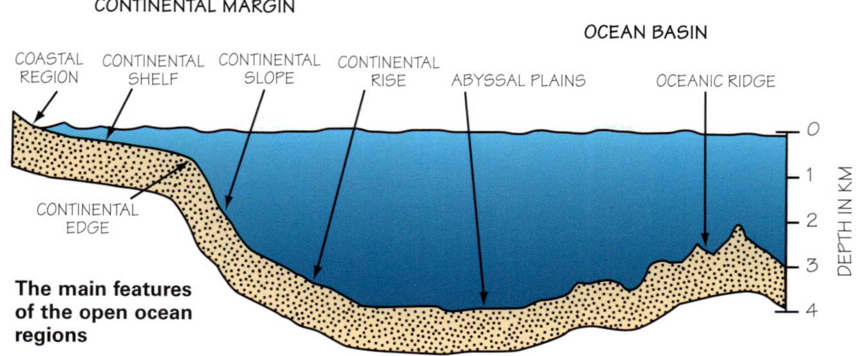

The main features of the open ocean regions

Continental Slope

Where the edges of the continents touch the sea, the geologic features are different. This area includes the continental shelf and the continental slope. The shallow sea bottom just offshore forms the *continental shelf*. This gradually sloping and shallow area contains water that is at most only a couple of hundred feet deep.

In some places, the continental shelf is only a few miles wide; in others, it can be a hundred miles or more. Beyond the continental shelf, the ocean depth drops off quickly. This slope, leading into deepwater, is called *the continental slope*. Occasionally, deep V-shaped valleys cut into the hard rock of the continental slope. These are called *submarine canyons,* which may be cracks from earthquakes or gullies cut by ocean *currents*.

Great currents heavy with sand and soil flow down the continental slope and deposit the material on the large flat areas of the ocean floor called the *abyssal plains*. These *turbidity currents* drop sediments rich in minerals and decomposing organic matter across most of the seabed, except on the mid-ocean ridges. The abyssal plains are larger in the Atlantic Ocean than in the Pacific because many of the world's major sediment carrying rivers such as the Mississippi and Amazon empty into the Atlantic. Also, large oceanic trenches scattered in the Pacific Ocean trap sediments. These trenches, found along the edge of ocean basins, are long, narrow, steep-sided depressions in the seabed. They contain the deepest parts of the ocean.

Some solitary undersea mountains called *seamounts* rise several thousand feet from the ocean floor. Their cone-shaped peaks remain submerged unless the water level drops. During the last Ice Age, much of the ocean's water froze to form vast glaciers, and the ocean's level fell much lower than what it is today. Wave action eroded and flattened the tops of exposed seamount peaks and resulted in what are called *guyots.* When the glaciers melted, the ocean levels rose and covered up those guyots.

Topographic features of the ocean floor

The Formation of Islands

The sea contains thousands of oceanic and continental islands. Geological disturbances beneath the water such as earthquakes and volcanoes create oceanic islands. Sometimes the peaks of giant underwater volcanoes reach the surface to form volcanic islands. The island of Hawaii is actually the top of a seamount. Where two plates of Earth's crust come together, volcanic islands develop in long, narrow, curved chains such as the Aleutian Islands by the Alaska Peninsula. The magma rising through the volcanoes helps to enlarge the islands in the arc.

Continental islands once joined the nearby continent. The connecting land gradually disappeared because of erosion or flooding. Along low sandy coasts, such as the eastern coast of the United States, long strips of sand called *barrier beaches* are separated from the shore by *lagoons*. *Barrier islands* are broader barrier beaches. These are separated from the mainland by narrow water passages called sounds.

During severe storms, barrier islands help protect the coastline.

Mountain Heights, Ocean Depths

Highest mountain on land: Mount Everest—29,035 feet (8,850 meters)
Highest mountain on Earth: Mauna Kea—33,476 feet (10,203 meters) from its base on the ocean floor. This mountain rises 13,796 feet (4,205 meters) above sea level, but almost 60 percent of its full height is below the surface.
Greatest known ocean depth: Challenger Deep—approximately 36,000 feet (10,973 meters)

Earth and the Sea

When you think of the ocean, you probably think of motion because the ocean is always moving. As a free liquid on a spinning planet, it is constantly tugged and pushed by forces near and far. Much of the ocean's motion takes the form of waves and tides.

Waves

The sea surface heaves and sighs as waves rise and fall. From earthquakes to ship wakes (waves created by boats), many forces create ocean waves. However, the most common force is wind. As wind passes over the water, it pushes on the ocean's surface, causing it to vibrate. That vibration creates a disturbance or ripple on the ocean surface. The strength of the wind, the fetch (uninterrupted distance the wind blows), and the duration of the gust determine how big the ripples become. During severe storms, the ripples can grow to waves 50 feet high.

A wave has several distinct parts. The crest is the portion above the still water line and highest point on a wave; the *trough,* or valley between two waves, is the lowest point.

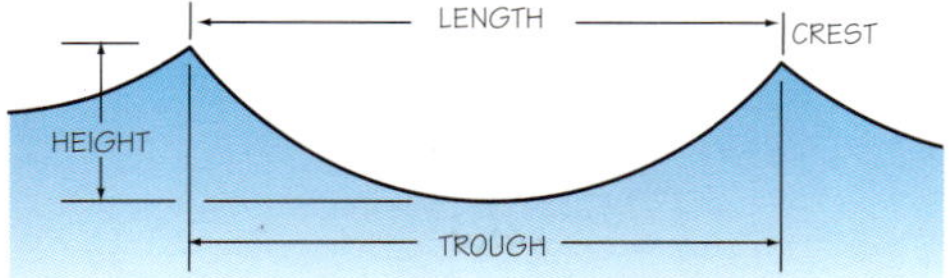

On beaches with gentle slopes, spilling waves break far from the shore, and the surf gently cascades down the front of the wave as it comes in.

The horizontal distance between the crests or troughs of two waves is called the wavelength. The vertical distance between the crest and the trough is the wave height. The wave period measures the period of time between two waves. You can determine the wave period by picking a point, say a rock or pier or buoy, and counting the seconds it takes for two waves to pass by. There—you're an oceanographer already!

Long, far-apart waves in the open ocean are known as *swells.* They travel faster than locally generated waves or *chop.* As a swell approaches shallow water, the ocean floor begins to affect the wave's shape and speed. Wave height increases and the crests become more peaked. As the steepness of the wave increases, the forward speed of the crest becomes faster than the speed of the wave, and the wave breaks.

Waves that break into foam are called breakers. Ocean swells breaking on the shore are called surf. There are different types of breakers; the most common are surging, plunging, and spilling waves. Where the beach slope is steep, surging breakers roll in and hardly break at all. On less steep beaches, plunging breakers curl over in a tube shape and finally break on the beach. These waves are great for surfing!

In deep water, even though a wave may be passing through, the surface may hardly move. That is because what is actually moving forward in a wave is energy passed *through* water, not water itself. If you watch a boat floating on water, you will notice that it goes up and down with the passing waves. The wave form moves along the surface of the water, but the boat stays in place.

Storm Surges, Tsunamis, and Tidal Bores

The strong winds of a hurricane or storm push seawater toward the shore. This advancing water may combine with normal tides to create a *storm surge,* which can increase the tide level 15 feet or higher. In addition, wind-driven waves roll in on top of the storm

In 2004, a powerful earthquake in the Indian Ocean caused a tsunami to hit island nations and coastal areas all over the region, from Southeast Asia to Africa. It resulted in mass destruction and more than 225,000 deaths.

surge, adding to the destructive power. Storm surges can cause severe coastal flooding, especially if the storm surge happens at high tide. Because so much of the U.S. population lives on the East and Gulf Coasts, many in locations just above sea level, the danger from storm surge is tremendous.

Occasionally, underwater disturbances such as volcanic eruptions, earthquakes, or landslides create monster waves called *tsunamis*. Reaching heights of 120 feet (37 meters) or more, tsunamis are the most dramatic and destructive of waves. The larger the underwater disturbance, the larger the tsunami. They have been called tidal waves, but their formation has nothing to do with the tides.

Tidal bores can push back rivers feeding into the inlet, making the rivers appear to run *backward*. A precise combination of conditions must occur for a tidal bore to form, so this phenomenon takes place in only a few places around the world, such as Canada's Bay of Fundy and the Amazon River in South America.

In the open ocean, tsunamis are hard to spot. Their long wavelengths mask their monstrous size, but like smaller waves, tsunamis change when they enter shallow water. Their wavelength shortens, and their crests rise to their full height. The strength of the underwater disturbance, the tsunami's wavelength, and the shape of the coastline all contribute to the tsunami's height and destructiveness.

Tidal bores are waves or walls of water that race up an inlet as the tide comes in. While not completely understood, tidal bores usually occur in V-shaped inlets that shallow up along their length. Wider at the opening and shallower at the head, these inlets force incoming water to collect in the middle. A wall of water then rushes up the inlet.

Rocky headlands are harder to erode than sandy beaches. Thus, headlands jut out into the sea, and sandy beaches curve away from the sea.

Waves and Coastal Formation

Waves form and shape coastlines. Wave erosion creates some of the world's most spectacular landforms, including sea caves, wave-cut notches, and coastal cliffs. Often it smoothes sandy beaches and forms barrier islands like North Carolina's Outer Banks.

Breaking waves can deposit or carry away sand and soil. This is known as deposition or erosion of sediment. As waves batter the coast, they erode and grind away the shore. Rocks and cliffs undercut by wave action fall into the sea and are ground and weathered into sand. The coastline's resistance to erosion determines its shape.

The ocean constantly reshapes beaches and replenishes the sand. Sediment deposits move along the seashore as every wave hits. Since the wind blows from different directions, waves rarely approach the coast head-on. As waves strike the shore, the *swash* (landward movement of water) carries sand to the beach at an angle. The *backwash* (seaward movement of water) returns straight out to the

Powerful water action carves steep and rugged seashores like this California coastline.

All waves use up their energy at the shore. Waves may break farther seaward on sandbars or reefs.

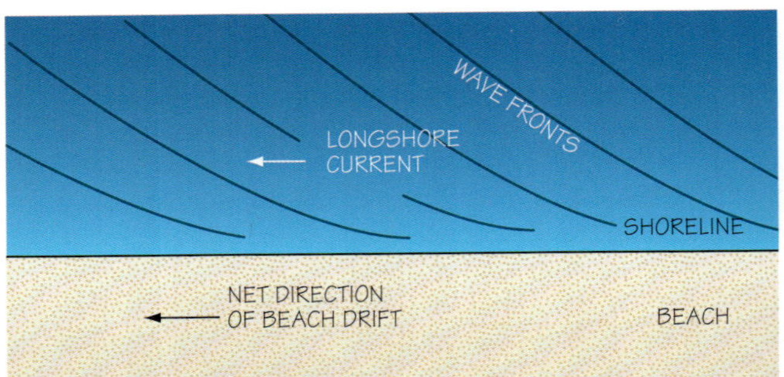

Formation of longshore current

ocean. Any sand carried by a wave that is not left on the beach is carried to the ocean by the backwash. There it settles on the seabed until another landward wave deposits it on the beach. This movement of sediment down the beach is called *beach drift.*

When waves hit the coast, some of the water flows along the beach, creating a longshore or littoral current. The current moves beach sediment in the water, a movement known as *longshore drift.* The combined movement of sediment via longshore drift and beach drift is called *littoral drift.*

The strength of the *longshore current* increases as the size of the waves and the approach angle increase. When the current grows strong enough to overcome the force of incoming waves, the water will flow seaward in a riptide, or *rip current.* A rip current can carry large amounts of sand and sediment away from the beach. If incoming waves do not return the sand, the beach gradually will wear away.

Rip Currents

Rip currents can carry a swimmer many yards offshore. Weak swimmers may panic and need help before they exhaust themselves trying to swim to shore against the current. To escape the grip of a rip current, the person should swim across the current, parallel to the beach and, when clear, swim for shore.

Given enough time, wave erosion will create a smooth coastline. How quickly the sea erodes a shoreline also depends on the amount of energy released by the waves as they approach the coast or shore.

The coasts you see today are the result of millions of years of geological evolution. People also have affected the shape of the shorelines by damming inland rivers and building barriers and other structures in the ocean. Dams on inland rivers diminish the amount of sand streaming into the sea. *Breakwaters* are structures built parallel to the shore to break the action of waves or to provide a calm harbor for boats. Groins angle away from the shoreline. *Jetties,* or piers, jut out perpendicular to the shore. These structures protect beaches by altering sediment deposits caused by inshore currents. Piers also jut out perpendicular to beaches and provide landings for vessels.

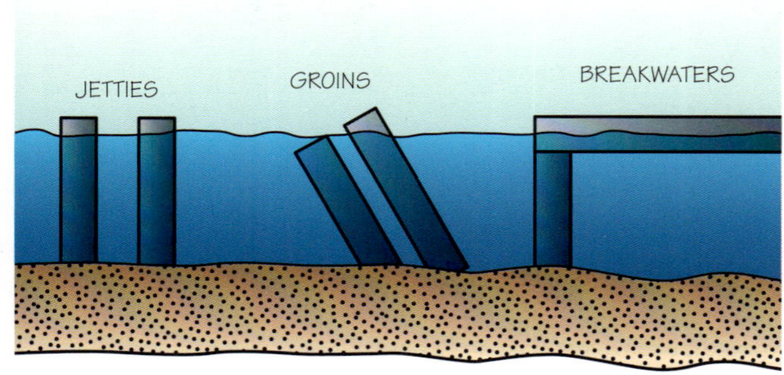

Artificial structures designed to break wave action block the migration of sand and sediment, causing some beaches to grow while others erode.

Not all ocean waves occur on the surface of the water. Internal waves occur within the ocean, between layers of water of different densities (imagine mixing oil with water), often when a tide containing these layers runs into some disruptive topographical feature on the ocean floor. The resulting waves can reach heights above 300 feet (91 meters), much higher than typical surface waves. Internal waves can influence the amount and diversity of nutrients available and the water temperature in an area. They can also hinder the monitoring of underwater environments and endanger submersibles.

On most ocean shores, high tide occurs regularly every 12 hours, 25 minutes. This means each new tide—high or low—occurs a little later each day.

A coastline's topography depends on the types of waves that hit the shore, the height of the tides, and the composition of the sediment and sand deposited on the surface. Coastal cliffs or steep banks may descend to dunes. The high point above the beach is the *berm,* a ridge formed by storm waves where seashells collect. The portion of the shore between the high-tide mark, usually a line of debris and seaweed, and the low-tide mark is the *foreshore.* If you wade ankle-deep into the sea at low tide, you will walk along the *low-tide terrace,* formed by the leveling action of low-tide waves. Further into the sea, you might find a longshore trough of water flowing parallel to the shore.

On the far side of the trough may be a *sandbar* or offshore bar. Sandbars are submerged or partially exposed humps of sand or coarse sediment built by the wave action of tides and currents. They frequently form in the heavy surf of the storm seasons but often are hidden in the deeper water. Breaking waves erode the tops of offshore bars.

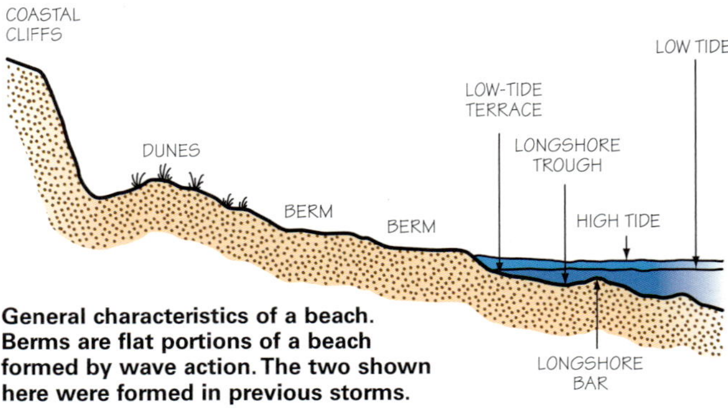

General characteristics of a beach. Berms are flat portions of a beach formed by wave action. The two shown here were formed in previous storms.

Beach waves excavate the longshore trough, and the swash deposits sand on the beach. Backwash and riptides move sand seaward to form sandbars, which may migrate shoreward in gentle seas and seaward in high seas. Landward waves add more sediment to the sand mass as they approach the shore.

Tides

The gravitational pull of the moon (and, to a smaller degree, the sun) on the sea causes ocean tides. In most parts of Earth, this pull produces two high tides and two low tides each day. As Earth rotates beneath the bulging waters, a high tide occurs, then a low tide, then another high tide and another low tide. The tilt of the moon's orbit gives the two daily high tides and the two daily low tides different heights.

Upwelling

Because the oceans have so much mass, they resist moving. Thus, only winds blowing over the water for long distances and for long periods of time are capable of generating ocean currents. The Gulf Stream is a warm current that flows from the Caribbean toward northern Europe on the western side of the Atlantic Ocean. Winds in the central part of the Atlantic drive this current. Along the eastern sides of the Atlantic, winds blowing in the direction of the equator push surface water offshore replacing colder, deeper ocean water. This process is called oceanic *upwelling,* and the nutrient-rich water that rises to the surface with the cold water helps support the abundant marine life of the Atlantic.

Our Liquid Planet

The most common substance on Earth is water. It also is very unique. No other substance on Earth acts or reacts like water.

Properties of Water

Water is colorless, tasteless, odorless, and wet. It serves as a cleaning agent, a heat absorber, a sound transmitter, a shaper of shorelines, and the medium upon which ships set sail and hurricanes brew.

At Earth's normal temperatures, only water can exist as a solid, a liquid, or a gas—ice, water, or water vapor. Water molecules are always moving; whether water appears as a solid, liquid, or gas depends on how fast they move. Ice molecules remain relatively distant and motionless. Liquid water molecules lie close together and move about. Water vapor molecules move rapidly and collide. Since water molecules are always moving, those at the surface break free of those below and enter the air as vapor. This process is called evaporation.

As most substances grow colder they contract (become smaller). But water is highly unusual. It contracts only until its temperature reaches 39 degrees, then it expands! For this reason, a can of soda left in the freezer will burst. Similarly, when ice forms on the sea, it floats. If water contracted when it froze, ice would sink and pile up on the ocean bottom. Summer's warmth could not reach deep enough to melt it so the seas would gradually freeze, killing all life on Earth. Thank goodness water is the way it is!

On Earth, water exists primarily as liquid. No other common substance stays liquid at room temperature. Between 32°F (its freezing point) and 212°F (its boiling point), water remains liquid.

A drop of water consists of many millions of tiny particles called molecules. Each molecule consists of even smaller particles called atoms. Two atoms of hydrogen combine with one atom of oxygen to form water—H_20.

Water has a great variety of characteristics. Among them:

Water can hold considerable heat. The ability to hold heat is called heat capacity. Only ammonia has a greater capacity to hold heat than water.

Water has surface tension. Surface molecules cling together so tightly that water can support objects heavier than itself. For example, insects may walk on water and pine needles may float.

Water is a solvent. Over time, ocean waves can dissolve even massive ocean cliffs. Water also dissolves nutrients on land that eventually find their way to rivers and into the sea. Inside an animal's body, water also helps dissolve food and carry it to the animal's cells. Because so many substances will dissolve in water, pure water cannot be found in nature.

If you want to make your own ocean water, you will need to add 35 parts of salt to 965 parts of water. That may not seem like much salt, but mix up a batch and have a taste!

Ocean Salinity

About 97 percent of all Earth's water is salty. What makes the ocean salty? It is no surprise that the main compound giving seawater its salty taste is the same salt we use on French fries—sodium chloride. But seawater also contains other salts such as magnesium chloride, magnesium sulfate, and calcium sulfate.

The amount of salts dissolved in ocean water is called salinity. Scientists use salinometers to measure the salt content of water. Salinity is measured in units called parts per thousand, or ppt. The average salinity of the world's oceans is 35 ppt.

Some areas of the oceans are saltier than others. Factors that influence seawater's saltiness include rate of evaporation, amount of rainfall, and how many rivers and streams pour into the sea nearby. Over time, rivers and streams carry great quantities of sediments and salts into the sea. On the other hand, rivers can also dilute the sea with freshwater. Some weather conditions such as hurricanes or tropical storms affect salinity because winds can sprinkle salts over water.

Temperature affects the amount of salt that water can hold. Warm water holds more dissolved solids, including salts, than cool water. Areas around the equator have warm temperatures and high evaporation. Few large rivers pour into the sea at the equator. Thus, seas around Earth's middle tend to be saltier than other oceans. Other areas, like the Gulf of Alaska, have a low rate of evaporation, a high precipitation rate, and a large number of freshwater rivers dumping their water into the ocean, lowering the average salinity.

How did all that salt get into the oceans? When oceans first formed on Earth, they were entirely freshwater. But for the several billion years since, continuous steady erosion of lands and mountains has carried salts and minerals to the sea. Gradually, the ocean's salinity has increased. Evaporation of seawater concentrated the salts even more. When seawater evaporates, the salt stays behind and the water becomes more saline. Evaporated seawater forms into clouds that rain down on the land again, picking up more salts and minerals, and flowing once more to the sea. Again, some of that water evaporates and leaves more salts behind, slowly increasing the ocean's salinity.

The oceans act as gigantic heat distributors, keeping the cold areas of Earth warmer and the warm areas colder. Without the oceans, the deserts would enlarge and the polar caps would contract.

Ocean Temperature

Ocean temperatures vary from the warm seas at the equator to the bone-chilling waters of the Arctic and Southern Oceans.

Oceanographers use an electronic instrument known as an STD (salinity, temperature, depth) to measure ocean temperatures within 4,000 feet (1,219 meters) of the surface. In the mid-1970s, the National Oceanic and Atmospheric Administration started using satellites to measure ocean temperatures. With a single pass of a satellite, scientists could get sea surface temperature data from New England to Florida. Over the course of a few days, they could record the temperatures of all oceans worldwide. Satellite technology has certain limitations in that it records only the temperature in the top inch or so of the ocean surface and is limited by cloud cover.

Iceberg

Ocean Density

Ocean density is the weight of seawater divided by the amount of space it occupies. Factors affecting seawater density include temperature, salinity, and pressure. Oceanographers express the density of seawater in grams per cubic centimeter. Seawater density increases as temperature decreases. Cold, salty water is much denser than warm, fresher water and will sink below the less-dense layer. Varying densities of seawater can create *deep ocean currents* and internal waves.

Density increases as pressure increases. Water weighs a lot; just fill two buckets full of water and carry them a hundred yards to demonstrate. Imagine the weight of a mile or two of water over your head. All that weight pushes down on the water deep in the sea and makes it denser. The average depth of the ocean is more than 12,000 feet (3,658 meters), which is more than 2 miles. So if you are feeling pressured, imagine the pressure deep-sea creatures must feel!

Gifts From the Sea

The ocean provides what people need: food, energy, minerals, medicine, and—of course—water.

The worldwide commercial fish and shellfish catch exceeds 200 billion pounds annually, most from waters near the coasts. Most fish and shellfish are harvested directly for food. Processors use the rest to make products such as fish oil and fishmeal to feed livestock and pets, and for fertilizer.

To support the high demand for fish, hatcheries produce salmon and other fry for ocean release. Fish farming, also called *aquaculture* or *mariculture,* produces fish, shellfish, and seaweeds near ocean shores.

The ocean is a source for energy. Offshore wells around the world tap deposits of oil and gas beneath the seafloor. Currently, these wells produce about 25 percent of the world's oil and about 20 percent of the world's gas. Ocean tides also provide energy. Tidal power facilities use the rise and fall of the tides to help produce electricity.

At undersea plate boundaries, vast accumulations of minerals form. Deposits near the *hydrothermal vents* contain copper, iron, and zinc. In certain areas, huge quantities of manganese collect on the ocean floor in lumps called *nodules*. Some undersea mining has begun, and large-scale undersea mining holds promise for the future.

Many forms of marine life contribute to modern medicine. Red alga provides an anticoagulant that keeps blood from clotting. One species of marine snail produces a substance used in muscle relaxants. Giant nerve cells from lobsters, squids, and marine worms help researchers learn more about nerve functions in people.

Seawater provides an inexhaustible supply of water. However, the salt must be removed before drinking it, or it will cause dehydration. This process—called desalinization—is costly, but it assures us that we can always get freshwater if we live near the ocean.

Know somebody losing his hair? Fish protein may be the cure for baldness.

Medicines and Chemicals From the Sea

The sea supplies a surprising number of modern medicines and chemicals. In fact, scientists scour the seas for medicines that may work better than those we now use. Fish oils reverse symptoms of arthritis and heart disease. Corals provide material for bone replacements. A polymer extracted from shellfish strengthens paper, improves cosmetics, stiffens hair gels, and helps prevent scarring. The next time you visit the beach, remember that the shells you see may turn out to be the wonder treatments of tomorrow.

Researchers use crustaceans—including lobsters—to learn more about how nerves function.

The Oceans and the Atmosphere

Two great systems envelop Earth—one is the ocean and the other is the atmosphere, or air. Both are constantly in motion, driven by the sun's energy and pulled by gravity. Each completely interacts with the other, giving and taking moisture, heat, and energy. Together, the oceans and the atmosphere affect climate and weather patterns around the world.

A Global Heat Absorber

The sun is Earth's main source of energy. The oceans, covering more than 70 percent of the planet's surface and darker than the continents, absorb roughly half of the solar (sun) radiation that strikes Earth. They store heat better than air and land do. Warm seawater mixes with cool seawater, thereby holding the heat energy.

Winds blowing over the warm ocean surface remove water vapor and heat. When the vapor condenses and falls as rain or snow, the heat energy released into the atmosphere causes the air to warm. As air warms, it rises. Then cold air flows to replace it. Sunlight heats the air unevenly. Air at the equator receives more sunlight and gets hotter than air at the poles. As hot air rises at the equator, colder air from the North and South poles rushes toward the equator to replace it.

Because Earth rotates, air does not flow in a north-south path, but is twisted. In the northern hemisphere, currents of air move clockwise. In the southern hemisphere, currents of air move counterclockwise. This phenomenon is called the Coriolis effect.

Great wind circulation systems form. Close to the equator, trade winds blow from the east. In the temperate zone, steady winds called *westerlies* blow from the west.

Winds blowing across the ocean's surface create currents. When the trade winds and westerlies blow across the oceans, they cause the currents in each ocean basin to move in a circular pattern.

Both surface and deepwater currents affect the world's climate by moving warm air from the tropics toward the poles, and cold air from the poles toward the tropics. For example, the Gulf Stream carries warm waters to northwestern Europe and Great Britain. As a result, London experiences much milder winters than New York City, even though Great Britain is situated at a higher latitude. As oceans circulate heat, they regulate Earth's temperatures.

Deep Ocean Currents

Latitude affects the temperature of ocean water. Cold winds blowing across the ocean at high latitudes (far in the north and far in the south) cool and evaporate the water. If the water is cold enough, sea ice will form. Because salts are left behind when sea ice forms, the cold water becomes denser and sinks deep into the ocean. The sinking and spreading of cold water is known as *thermohaline circulation,* or deep ocean currents.

Scientists have known about this very cold water in the ocean basins (even at the tropics) for a long time. There is much more of this deepwater than there is of surface water. While deep ocean currents are not as strong as those of the surface water, they are nonetheless important in *ocean mixing,* a way for seawater carrying dissolved gases and important nutrients to mix with nutrient-poor seawater.

So what does the atmosphere have to do with deep ocean currents? When clouds—which are part of the atmosphere—block the sun's rays from the ocean, the ocean cools. These same clouds also might bring rain, which is fresh water. The freshwater reduces the salinity of the ocean. Winds blow in and evaporate some of the ocean water, and when the water vapor rises, the salt stays behind. Temperature, precipitation, and wind thus affect density, which affects deep ocean currents.

Despite their incomplete understanding of the effects of methane, natural trace gases, and industrial pollutants, many scientists believe that the rise in global temperatures in recent years is a result of the greenhouse effect.

The Sea and the Greenhouse Effect

The ocean absorbs and dissolves various gases from the atmosphere, including oxygen, nitrogen, and carbon dioxide. It can hold enormous quantities of dissolved gases because it has so much dense cold water. The oceans are the main reservoir of dissolved carbon dioxide, an important *greenhouse gas.*

Greenhouse gases are those that affect Earth's surface temperature. These gases include carbon dioxide, methane, and water vapor. Many people overlook water vapor as a greenhouse gas, but it is the major reason why humid regions in the world experience less cooling at night than do dry regions. Greenhouse gases trap and hold the sun's heat. Without these gases, the average temperature of Earth would drop below the freezing point of water.

Some environmental scientists are concerned that changes in the atmosphere may be affected by or may be intensified by human activities and could cause Earth's surface to warm to a dangerous degree. This effect is called global warming. These scientists say even a limited rise in average surface temperature could lead to a partial melting of the polar ice caps and glaciers, which would cause a major rise in sea level, along with other severe environmental disturbances. They believe global warming could cause significant changes to habitats and weather patterns, which could endanger plants and animals.

Some scientists say that in recent decades, there has been a global increase in atmospheric carbon dioxide because people have been burning fossil fuels (coal, oil, and natural gas). If the present global climate remains constant, they say the increase in carbon dioxide from all this burning could raise the average temperature at Earth's surface. Warm air can contain more water than cooler air can, so a warmer atmosphere will hold more water vapor. The scientists believe this cyclical process might continue to raise the temperature at Earth's surface.

We still have a lot to learn about the role of oceans in the atmospheric carbon cycle. While about half of the carbon dioxide produced by burning fossil fuels and by deforestation is dissolved in the ocean—slowing global warming—temperatures have, on the average, been rising. Moreover, the excess carbon dioxide entering the ocean lowers its pH, and the resulting acidification threatens organisms with calcium carbonate shells and skeletons and coral reefs.

La Niña (lah NEEN yah), El Niño's cold counterpart, has similarly important effects on weather patterns. It is associated with drought in the southern United States and enhanced hurricane activity in the North Atlantic Ocean, for instance.

El Niño

El Niño (ehl NEEN yoh) is a warm current in the Pacific Ocean that flows southward along South America's west coast. It usually lasts nine to 12 months and returns on average every four years. Normally deep, cold Pacific waters well up to the surface off the Peruvian and Ecuadorian coast, bringing up nutrients and chilling surface waters and causing climatic changes worldwide. A powerful El Niño in 1982 and 1983 caused severe drought in Australia and Indonesia and an unusually large number of storms in California. During the 1997–98 El Niño, heavy rains caused flooding in Peru, Ecuador, and Argentina; monsoons and cyclones struck Madagascar; and severe storms pelted Texas and the Southeast. More recent El Niño events in 2003 and 2006 were not as devastating.

El Niño is the focus of many studies. Scientists believe El Niño is related to a shift in air movements over the tropical Pacific Ocean. Changes in wind direction cause changes in the circulation and temperature of the ocean, which in turn further disrupt air movements and ocean currents. The location of warm and cold pools of surface ocean water may also influence the location, movement, and intensity of storm systems in the atmosphere.

Our Living Sea

A rich assortment of living things calls the ocean home. Big and small, long and short, multicolored and drab, active and inactive, sea life ranges from microscopic plankton to the largest creatures on Earth. Some organisms even light up, like ghostly neon lights.

Zones of Life

The sea is so vast that scientists have divided it into zones. The *pelagic zone* includes all the environments of the ocean above the bottom, or living in open oceans or seas rather than in waters next to land or inland waters. It is divided into an inshore *neritic* zone (the zone of shallow water adjoining the seacoast) and the open-sea *oceanic* zone. The boundary between them occurs at the edge of the continental shelf. The oceanic zone is divided further according to how deep sunlight penetrates.

Planktonic plants occur only in the neritic and epipelagic zones but provide food to animals living in the water and on the bottom. Open-ocean life forms are called *pelagic;* bottom-dwelling life forms are called *benthic.* The *benthonic zone* is subdivided into three bottom zones: the littoral, bathyal, and abyssal.

Plants grow only in the sunlit zone, the only ocean layer that absorbs enough sunlight for photosynthesis. Animals live in all the oceanic zones, although because of the availability of food, more of them are found near the ocean's surface. The sunlit zone is very shallow compared to the bathyal or abyssal zones.

- The **epipelagic zone**—more commonly called the sunlit zone—is the top section of the ocean. This is the only section shallow enough for sunlight to penetrate, allowing photosynthesis (discussed later) to take place.
- The **mesopelagic zone**—also called the twilight zone—is dimly lit. There are no plants in this zone because it does not get enough sunlight to generate photosynthesis.
- Some call the **bathypelagic zone** the midnight zone. No sunlight penetrates this area of the ocean because it's too deep—about 2 1/2 miles down. The only light that this deep ocean layer gets comes from bioluminescent marine life.
- In the **abyssal zone,** the ocean's lowest layer, it is pitch-black and—since it is unaffected by weather—calm. The temperature is nearly freezing at this depth, around 6,600 to 20,000 feet (2,012 to 6,096 meters) below the ocean's surface.
- The **hadal zone** comprises the ocean's deepest waters, found in its narrow trench walls and floors—some deeper than 6 miles (nearly 10 kilometers).

Life Near the Surface

Food is most abundant in the sunlit zone. There, the majority of ocean life finds the food sources needed to survive—mostly in the form of microscopic algae.

The Intertidal Zone and Intertidal Invertebrates

Abundant and varied plant and animal life thrive in intertidal communities between the high and low tides of marine coasts. Factors such as the type of rock, type of sand or soil, water temperature, protection from waves, and the interactions between organisms determine what an intertidal community is like. Intertidal communities are rich in life, especially in invertebrates—animals such as clams, mussels, starfish, and others that do not have a backbone or spinal column.

The invertebrate group is one of two general categories of animals. The other group, vertebrates, includes those animals that do have backbones (fishes, amphibians, reptiles, birds, and mammals).

Invertebrates occupy all habitats—even deep-sea trenches—and are found in all types of sea-bottom sediments, from soft oozes to rocky bottoms. Swimming invertebrates survive at all depths and include forms developed to live in the sunless waters of the deep sea as well as near the surface. Invertebrates exist in fresh, brackish (slightly salty), and marine environments. Some specialized forms also thrive in extremely salty seawater such as lagoons along tropical coasts and pools high in the intertidal zone.

Organisms in the *sub tide zone* are fragile and cannot tolerate much exposure to the air or sun.

Marine invertebrates include the giant squid, which can measure up to 65 feet (20 meters) long and weigh 2 tons or more. Since invertebrate simply means "without a backbone," in number of species, this category constitutes almost the entire animal kingdom.

Splash and the Upper Intertidal Zones

In the intertidal zone, the area highest above the waves is the *splash zone,* which is just reached by the ocean's salty spray. Animals living in this zone, such as shore crabs and sand fleas, are primarily adapted to life above the waves. Only a few hardy animals live in splash pools, which dry up in summer and flood with freshwater runoff in winter.

Below this zone lies the *upper intertidal zone* with its scattered covering of green and brown seaweed. Here, snails and limpets scour rocks in search of microscopic algae. Barnacles cover the rocks except where predators and winter storms have cleared them away. Because barnacles cement themselves to the rocks and close when the tide is out, they are well-adapted to this zone.

Mid- and Low Intertidal Zones

Below these two zones, in the mid-intertidal zone, a broad band of mussels often forms a bed several inches thick. The common starfish continually feeds on mussels, which would otherwise abound throughout all the available mid-intertidal space. Many worms, snails, and crabs live within the mussel bed. The mid-intertidal zone is covered and uncovered twice a day by the tides. Animals in this zone have adapted to being immersed in air and seawater.

The diverse plants and animals of the low intertidal zone include red algae and large *kelp,* which clings to the rocky bottom with *holdfasts.* The kelp's large leaves move with the waves, and the kelp beds protect the sea urchins, worms, snapping shrimp, and porcelain crabs preyed upon by the giant sunflower starfish.

Sandy beaches, while fun for sunbathers, provide limited habitat for marine and shore animals. Those that are there, though, are often found in abundance, especially crabs, clams, and beach hoppers.

Coastal Marshes

The quiet waters of bays and river mouths are lined with grassy marshes. In these marshes, fiddler crabs undercut the banks while oyster banks fill the lower edges of the intertidal marshes, sheltering crabs and snails. Clams, shrimp, and worms burrow into mud flats for nutrition and protection. These animals eat by filtering the water at high tide or by scouring the mud bottom for tiny food particles. The holes they make provide habitat for shrimp, tiny crabs, and small fish.

Sediments in an estuary come from rivers and the ocean. As tidal currents push sand into an estuary, rivers drop sediments. Over time, estuaries fill with soil and disappear.

Estuaries

Estuaries are river valleys flooded by the sea. It's where fresh-water and seawater mix. When the sea level rises or the land subsides, the sea can cover parts of the coast. An estuary like Chesapeake Bay on the East Coast of the United States has an average depth of about 13 feet, despite the deep channel down its center. Estuaries provide a rich habitat for shellfish that live in salty water. In the inner bays of the estuary, less dense river water flows over denser seawater.

Beds of seaweed provide a habitat for many animals. Some seaweeds, such as *sea lettuce* (shown here) and *dulse,* are used as ingredients in human foods or as garden fertilizer.

Seaweeds

Seaweeds are algae that grow in the sea. Red and brown algae are the most common, although a few green algae also are included. Seaweeds usually attach themselves to rocks or to the ocean bottom. Mats of seaweed are often carried and left on a beach by high tides. Although seaweeds thrive at depths of up to 100 feet (30 meters), some red algae grow as deep as 600 feet (183 meters).

Coral Reefs

A *coral reef* is a rise or mound of *coral,* coral sands, or solid limestone at or below sea level. Coral may make up less than half of the reef; other organisms such as *mollusks* (snails or clams), *zooplankton,* and sponges form the rest. Finally, coralline algae commonly help to bind these organisms together to form the framework of the coral reef.

Coral reefs have three forms: *fringing reefs, barrier reefs,* and *atolls.* In tropical areas, fringing reefs form just offshore, separated from land by shallow water. In other areas, barrier reefs form far from land, separated by water more than 30 feet (9 meters) deep. Atolls, found far offshore, are rings of coral encircling a lagoon.

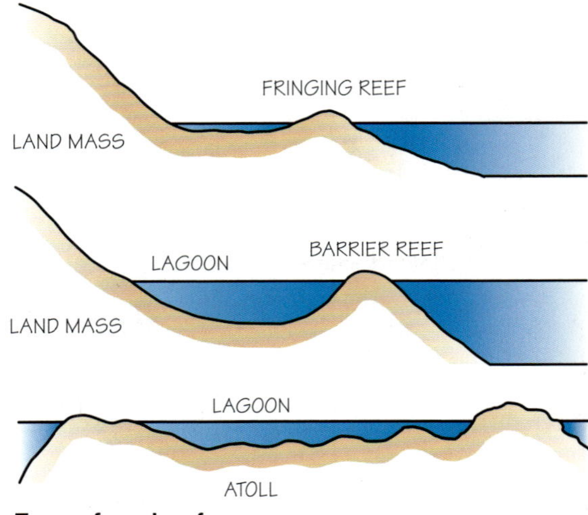

Types of coral reefs

How Coral Reefs Form

During his scientific expedition aboard the HMS *Beagle* from 1831 to 1836, Charles Darwin developed a theory about how coral reefs form. First, a volcano grows thousands of feet from the ocean floor to rise above the surface of the ocean. It becomes a volcanic island surrounded by shallow water. A shelf of coral extending from the shore forms a fringing reef around the volcano top. Gradually,

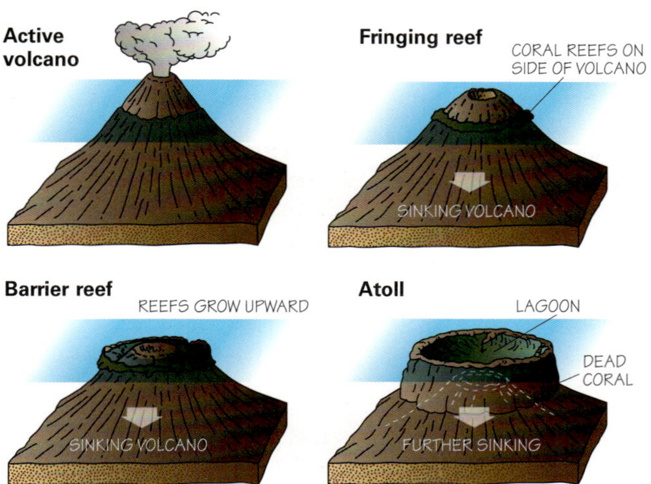

Darwin's theory of the formation of an atoll

This giant clam, *left,* was spotted at Australia's Great Barrier Reef, *right,* which is the world's largest coral reef.

the volcano sinks, but the coral continues to grow. As the volcano continues to sink, the reef becomes separated from the landmass by a lagoon. It is now a barrier reef. Finally, the volcano disappears far below the ocean's surface and leaves only an atoll, a ring of coral with a lagoon at its center.

Coral reefs are largely made of the skeletons of *colonial corals.* Not all corals form into reefs. Some live a solitary existence, like their relatives, the *sea anemones.* Temperature, water depth, salinity, and wave action all help determine the growth and health of corals.

> Coral reefs benefit from heavy wave action. Waves agitate the water, bring in food and oxygen, and remove sediment.

Tropical corals require waters warmer than 65°F. These reef corals rely upon algae for nutrients and oxygen. Algae require sunlight for photosynthesis. Tropical reefs, therefore, are found in water shallow and clear enough for sunlight to penetrate.

The framework of the coral reef is provided by coral and some minor organisms, then cemented by the coralline algae. The loose sediment used in this framework is made up of eroded reef rock, sand, gravel, and a small amount of silt and clay.

Cold-water reefs (also known as deepwater reefs), on the other hand, exist in waters between 39°F and 54°F, at depths that have limited or no sunlight. These reefs have not been studied nearly as much as tropical reefs, and much is still unknown about cold-water reefs. However, they are gaining more attention in ocean research, and they continue to be discovered and explored all over the world. Like their tropical counterparts, researchers have found them to be home to a wide variety of organisms.

Coral Reef Life

Tropical coral reefs are usually found in warm shallow waters on the eastern coasts of continents and around oceanic islands. Most of the world's tropical reefs are found in the Indian and Pacific Oceans and the Caribbean Sea. Cold-water reefs are found in colder, deeper waters.

Numerous examples have been found in the Atlantic Ocean, but they have also been found in the Indian and Pacific Oceans.

Coral reefs seem to have few plants, but reef-building corals often house microscopic single-celled algae called *zooxanthellae.* The zooxanthellae supply food through photosynthesis. The association between coral and zooxanthellae benefits both. The coral gives a place for the algae to form, and the algae provides nutrients for the coral. Also, corals feed upon plankton and organic debris.

Because some corals must spread out in search of light much like a tree grows branches, the association between coral and algae explains why many coral reefs grow like a forest. Corals spread out and compete for light just like plants on land do. Just as with plants on land, rapidly growing corals shade out slower growing colonies, which eventually die.

This coral shows the effects of bleaching.

Today, tropical and cold-water coral reefs in some parts of the world are in peril. When stressed by pollution or a change in temperature or sea quality, or for as yet undetermined reasons, some reef corals expel the algae, weaken, and die. This condition is known as "bleaching." Human activities, such as tourism and commercial fishing, also threaten the reefs. To make matters worse, the rate of growth—and therefore recovery—of coral reefs is slow.

Plankton

The term plankton comes from the Greek word *planktos,* which means "wanderer." Plankton includes marine plants and animals that drift with the currents. Because they drift, plankton differ from *nekton,* animals that actively swim or lie on or burrow into the seafloor, such as clams and worms. They also differ from marine plants such as large seaweeds. Plankton cells are seldom larger in diameter than a few tenths of an inch. Plankton forms the basis of all major ocean food chains.

Plankton is denser than water and tends to sink. However, because of their small size, long spines, shell extension, and the ability to float, these drifters have adapted in different ways to slow this sinking. Some planktonic animals rise toward the surface at night and then sink to deeper waters during the day. This *vertical migration* is probably tied to feeding strategies as well as to the avoidance of predators. When plankton dies, it sinks and contributes to the rich sediments at the bottom of the sea.

Phytoplankton

Plant plankton, or *phytoplankton,* consists of single-celled algae. Through photosynthesis, plankton supports the rest of all marine life. Photosynthesis is the process plants use to convert sunlight into food and release oxygen from carbon dioxide and water. The amount of available light and nutrients affects phytoplankton production.

Seawater absorbs sunlight. The deeper you go, the less light is available. The depths of the sea are pitch-black. Between the dark depths and the surface is the euphotic zone, the layer of seawater that receives enough sunlight for photosynthesis to occur and plants to grow.

Besides sunlight, phytoplankton needs nutrients to survive. Ocean mixing brings these nutrients up from the seafloor. Levels of plankton tend to remain low and constant year-round in the clear open ocean and tropical waters. Light can penetrate deeper there, but storms have trouble churning up nutrient-rich bottom sediments in deepwater. The most productive areas are where surface currents lead away from land and mix with deep currents, like the west coast of Ecuador and Peru and off western Africa.

Zooplankton

Planktonic animals, or *zooplankton,* are divided into two groups. *Holoplankton* spend their entire life cycle as plankton. These include small crustaceans such as crabs, krill (the principal food of baleen whales), and jellyfish. *Meroplankton* are plankton only during the larva stage (the earliest stage of an animal, before it changes and becomes an adult). These species, which include shrimp, barnacles, worms, and marine fish such as herring and anchovies, change dramatically as they grow. As many as 50,000 zooplankton may inhabit one gallon of seawater!

Marine birds include stilt-legged birds such as herons and egrets, terns and skimmers, gulls, cormorants, pelicans, frigate birds, pelagic birds (which spend nearly their entire lives beyond sight of shore), and penguins.

Ocean Food Chains

Flesh-eating animals are called carnivores, whereas plant-eating animals are called herbivores. Those that eat both plants and meat are called omnivores. In some way, all animals depend on photosynthesis to grow plants that help keep the food chain going.

As microscopic plants, phytoplankton claim the basic level of the ocean food chain. Zooplankton graze on these phytoplankton. Copepods, larvae, and other tiny animals filter these algal cells from the water. Larger zooplankton prey upon these zooplankton and then become prey to even larger zooplankton. Several levels of preying occur before a large fish such as a cod or tuna appears in the ocean food chain.

Marine Birds

Marine birds have adapted to the ocean. Many of them possess webbed feet. Others have glands that empty salt and excretory systems that conserve water. Where food is abundant, marine birds are as well. Each species has developed special features that enable them to feed on different foods, nest in different places, and remain active at different times of the day. This reduces competition for food among marine birds. Though marine birds feed on fish, squid, dead and floating fish, and animals at sea, they must return to shore to nest.

Some marine birds possess fatty deposits and thin, light bones and oil glands near their tails to waterproof their feathers. Others have air sacs in the thorax, abdomen, and long bones of their legs and wings to help them float. Many seabirds, such as penguins, use air trapped under the feathers to insulate their bodies. Diving birds exhale from air sacs and lungs, squeezing air from under their feathers, to push below the surface. Their heart rate slows when they dive.

Penguins have blubber under their skin and scalelike feathers to help keep them warm and survive harsh winters in the Antarctic, where the temperature can dip to -70°F.

Seabirds are vital to the marine food web, feeding on fish and adding droppings to water, fertilizing the sea and stimulating the growth of marine plants.

Greater numbers of fish occur near coasts, where the waters are rich in nutrients. There, sunlight warms the shallow waters, and upwellings from ocean depths mix with nutrients brought to sea by rivers, providing rich habitat.

Ocean Fish

Though fish are cold-blooded, they are vertebrates (they have backbones). But some fish are flat and lie on the bottom of the ocean like halibut or swim upright like sea horses. All fish have fins; virtually all have gills. Fishes range in size from tiny tropical fish such as the pygmy goby of the Philippines, with adults as small as your fingernail, to the whale shark, which grows to 60 feet long and weighs more than 20 tons.

Distribution of Ocean Fish. Fish inhabit the oceans from the shallowest shoals to crushing depths 6 miles (almost 10 kilometers) deep. They are found from the balmy tropics to the icy Antarctic, where the seawater is colder than ice. Twenty thousand species of fish swim the seas, outnumbering all the other vertebrate species combined.

Anatomy of Ocean Fish. Fish are composed of three classes: the cartilaginous-skeleton fishes, such as sharks and rays; the jawless fishes, such as sea lampreys; and the bony-skeleton fishes, which are all the rest.

Fish extract oxygen from water. Most fish have gills, but some have lungs or other ways to absorb oxygen, including through their skin. Breathing underwater is harder work than breathing above water. Water is dense and holds far less oxygen than air.

Scales protect the skin on a fish, but not all fish have them. Scales are colorless and vary greatly between species.

In the sea, fish lose water and absorb salts. To offset this loss of fluid, marine fishes drink seawater and produce very little urine. The drinking of seawater, however, means that fish accumulate more and more salt. This excess salt is eliminated through the anus along with wastes. Some wastes are excreted through the gills. Fish get rid of so much salt that their blood is much less salty than seawater.

The pectoral fins situated at the front of the body behind the gill openings allow fish to maneuver and, if need be, to hover in place.

Marine Mammals

Marine mammals are animals that spend all or most of their time in the ocean, where they find their food. Animals such as whales, dolphins, porpoises, seals, sea lions, walruses, dugongs, manatees, and sea otters are marine mammals. Like land mammals, they nurse their young with milk produced by their bodies. The polar bear also is considered a marine mammal because it is a proficient swimmer and hunts primarily on sea ice for ringed and bearded seals.

Whales

Whales are the largest marine mammals. Extant (still existing) whales include the baleen and toothed whales. The body of most whale species is shaped like a torpedo. It uses its front limbs mostly to steer and balance its large body. The hind limbs have only a few bones. Whales swim with help from the vertical movements of their dorsal fin and tail fin, which consists of a pair of horizontal lobes called flukes.

Baleen Whales. The baleen whale has no teeth and wouldn't need them, anyway, because it is primarily a plankton feeder. It gets most of its nourishment from tiny ocean organisms filtered by a row of fringed plates of *baleen,* or whalebone. The blue whale—at up to 100 feet long and with an estimated weight of more than 130 tons—is the largest living animal and the largest animal that has *ever* lived. Some other baleen whales include the humpback, gray, and fin whales.

Humpback whale

Toothed Whales. Toothed whales have sharp teeth, usually in both jaws. They are predators, feeding on fish and squid. These whales include the dolphin, sperm whale, narwhal, beluga, porpoise, and killer whale. Possibly because of Herman Melville's novel *Moby-Dick,* the best-known toothed whale is the sperm whale, which can grow to longer than 60 feet. The sperm whale, with its oblong head and narrow lower jaw, is an amazing predator, diving to great depths in search of squid. Other toothed whales include the following

- Narwhals are found along coasts and in rivers throughout the Arctic. They feed on fish, octopus, and crabs. Eskimos commonly hunt narwhals.
- Beluga whales inhabit the Arctic Ocean and adjacent seas, in both deep offshore and coastal waters. They may also enter rivers that empty into far north seas.
- Orca whales are found in all seas from the Arctic to the Antarctic. The largest of the dolphins, they attain a maximum length of about 31 feet and a weight of about 9 tons. The orca whale is black, with white on the underparts, above each eye, and on each flank. Orcas live in pods, or social groups, usually of a few to about 50 individuals. Orcas are sometimes called "killer whales" because they are aggressive hunters. Despite their nickname, these whales have not been known to harm humans in the wild.

Orca whale

Most small-toothed whales are dolphins. They have a beak-like snout and sharp, pointed teeth (like Flipper). People sometimes mistake porpoises for dolphins. Porpoises have a rounded snout, chisel-shaped teeth, and a triangular rather than hooked dorsal (top) fin, and they usually are smaller than dolphins.

Porpoise

THE DISTRIBUTION OF WHALES

Whales are found in all the oceans, from the tropics to the icy northern and southern latitudes. Many whales migrate. The migration paths of gray whales follow the longest round-trip migration of any mammal, nearly 15,000 miles. The blue whale, the sperm whale, and other whale species can be found in any ocean the world over.

WHALE BEHAVIOR

Although little scientific evidence pinpoints any consistency in the social behavior of whales, it is clear that most travel in schools, also called pods or gams. Whales are very social and seem to communicate through "songs" and other sounds they broadcast in the water. You may have seen whales propel themselves completely into the air—a motion called "breaching." While it has not yet been proven, some believe breaching is tied to the whale's mating and other social behaviors.

Seals, Sea Lions, and Walruses

Seals, also known as earless seals, are classified with walruses and the eared seals (sea lions and fur seals). Seals are especially numerous in the colder waters (above 40 degrees latitude) of both hemispheres, with concentrations in the polar regions. Monk seals live in the warmer waters of the Caribbean and Mediterranean Seas and Hawaii. The most numerous seal is the crab-eater, with a population of more than 14 million.

Its thick layer of fat, or blubber, insulates the seal from the cold, helps protect it from serious injuries, improves its ability to float, and serves as a source of stored energy.

The smallest seals are the ringed seals, which reach an average length of about 3.5 feet and an average weight of about 110 to 200 pounds. The largest are the elephant seals, which can grow to up to 21 feet long and weigh about 7,780 pounds. The Weddell seal can dive as deep as 2000 feet and remain underwater over an hour. Seals, sea lions, and walruses usually eat fish, mollusks, crustaceans, and squid. The leopard seal of Antarctica, however, eats warm-blooded animals—including penguins and other seals.

Sea lion

Life in the Deep Sea

All ocean waters beyond the continental shelf and below the level of light penetration comprise the deep sea. While its conditions are harsh, the deep sea is an enormous area—some 90 percent of the ocean is deep sea. Because of the pressures, cold, and darkness, the area is virtually inaccessible to humans, but it still harbors life.

The Deep-Sea Environment, Hydrothermal Vents, and Cold Seeps

Seawater weighs a lot. The deep sea experiences enormous pressures, and it is cold and lightless. At any one depth, the deep-sea environment is constant. The temperature stays the same, the salinity remains the same; it is a chilling, yet fascinating place. Surprisingly, except for the bottom of some deep oceanic trenches, these chilling deep-sea waters still contain enough oxygen to sustain life.

Though food is often scarce in the deep sea, food drops from the surface waters above or is brought in by ocean currents from coastal areas. However, certain creatures count on bacteria around hydrothermal vents and *cold seeps,* which are found at several locations in the Atlantic and Pacific Oceans.

In cold seeps, methane and sulfide-rich fluids seep from the ocean floor. The cold seeps support organisms similar to those supported by the hydrothermal vents. Unlike other marine animals, these organisms (many are new to science) do not depend on food sinking down because they can feed off the nutrients present in the cold seeps.

Hydrothermal vents are cracks in the ocean floor from which heated water continually rises. They may be the main way Earth's core loses heat. Although the vents occur in the abyssal zone, they support communities of unique species—including giant clams, limpets, mussels, and tube worms up to 10 feet long—that are dependent on sulfur-digesting bacteria for energy.

Deep-Sea Life Forms and Adaptations

Most plants cannot exist in the absence of light. However, the deep sea contains a wide range of animals that can survive in the great pressures. Most deep-sea fishes are much smaller than their shallow-water relatives. But deep-sea crustaceans often outsize those in shallower waters. Fishes of the shallower parts of the deep sea may have very large eyes to catch what light is available in the upper zones.

In some animal species, their glow is produced by bacteria rather than by the animals themselves. Colonies of bacteria produce light in anglerfish.

Bioluminescence

In the deeper zones, some species of both fishes and invertebrates, such as marine bacteria, crustaceans, fish, fungi, jellyfish, mollusks, protozoa, sponges, and worms, are able to produce light. The production of light by living organisms is called bioluminescence. Many deep-sea fish have bulblike organs on their bodies that may also attract mates or prey, or illuminate the search for them in the perpetual darkness.

Luminous single-celled organisms called dinoflagellates are the most common source of brilliant displays of light seen in the ocean. One species may tint the sea surface pink in the daytime and light it up at night. A ship plowing through tropic seas may produce a wake that glows eerily as millions of these organisms light up. Another source of light in tropical oceans is the luminous jellyfish.

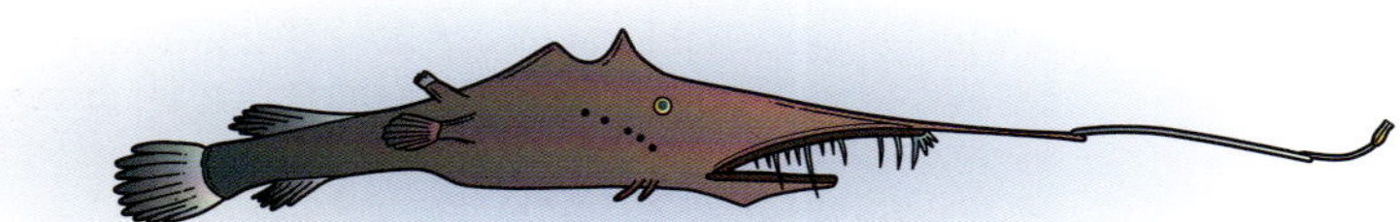

Deep-sea anglerfish have a rod on their heads with a "light" attached at the end. Interestingly, the female grows up to 3 feet long, while the full-size male measures only about 5 *inches.*

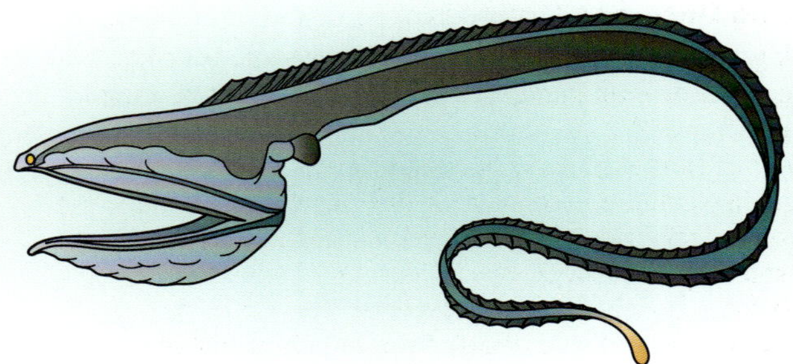

The deep-sea gulper grows up to 2 feet long, and up to a quarter of its length will be that of its mouth alone.

Adaptations of Deep-Sea Fishes

Since food in the deep sea is scarce, many deep-sea fishes have large mouths that enable them to swallow prey species larger and wider than themselves. Backward-curving teeth ensure that the prey does not escape. Many fishes also have guts that expand to digest meals larger than themselves.

Because of pressure effects, the body composition of fishes and invertebrates in the deep sea differs from that of shallow-water forms. The flesh of deep-sea animals is jellylike.

Studying Deep-Sea Organisms

Sampling and working with deep-sea organisms is not easy. Getting to the deepest part of the sea and being able to bring live organisms to the laboratory is extremely difficult.

Benthic (bottom) plants and animals are easier to study because they either do not move or they move slowly. The organisms of the bottom ooze are usually sea cucumbers, brittle stars, small crustaceans, worms, and mollusks.

Whereas most benthic organisms feed on materials drifting down, virtually all pelagic fishes and invertebrates are carnivores in the deep sea. In the upper layers, many of these animals migrate, moving toward the surface at night and returning to the depths during the day.

Deep-sea tube worms

Alvin, **perhaps the most active and successful research submersible**

Technology and the Sea

Humans have always depended on technology to explore the oceans. Lighthouses guide sailors away from reefs, and compasses, radio, and radar allow ships to chart their course. Today, Global Positioning System (GPS) satellites tell mariners their exact position. Sonar equipment, used to determine the shape and depth of the ocean floor, also locates submerged submarines and schools of fish.

Marine Research

Today's oceanographers explore the oceans using research ships, research submarines called submersibles, remote and autonomous sensing devices, and satellites. Using computers, they compile and analyze the data collected. They can project trends and create biological, geological, chemical, and atmospheric models (mathematical representations) critical to understanding the oceans.

***Aquarius,* an underwater laboratory in the Florida Keys, allows researchers to participate in 10-day missions such as studying the decline of coral reefs in the area.**

Orbiting satellites may monitor dozens of sensing devices at once, providing ongoing snapshots of ocean currents.

Oceanographers might be at sea for days or weeks studying marine life or ocean conditions. Many ships are laboratory-equipped; a few carry marine submersibles. Though modern ships are well outfitted, they often represent an inefficient way to conduct research since they can only be in one place at a time—often at great expense. Sometimes rough seas make for uncomfortable conditions.

Oceanographers may use underwater camera or video equipment. They take measurements using sonar, recording sound waves echoing off the ocean floor to better understand sea depth and seawater density. They may penetrate Earth's crust with sound to better understand undersea geology. Oceanographers collect seawater samples from various depths and measure the temperature, salinity, and other characteristics of the samples. Nets towed behind research ships gather samples of marine life for study.

Oceanographers use remote sensing devices to record data over time or in difficult locations. Some devices float or drift on ocean currents above or below the ocean's surface, providing important information about atmospheric pressure, water temperature, and ocean currents. Some are anchored to the ocean floor at a certain depth. Others record data over a long period

The *Seward Johnson I,* a research vessel, is prepared to deploy the submersible *Johnson Sea-Link.*

The *Johnson Sea-Link* is equipped with a variety of tools that scientists use to collect samples from the ocean depths.

of time, so oceanographers and technicians are involved in machine design and maintenance.

Some oceanographic vessels are equipped to drill into Earth's crust. Undersea rock, sediment, and core samples may help scientists better understand the age, composition, and development of the ocean floor.

Submersibles enable scientists to observe undersea life and geography that other instruments may miss. Some submersibles, such as the *Alvin* and *Turtle,* carry a human crew. Sometimes submersibles have carried divers deep into the ocean where they leave the sub to study the ocean bottom or to collect marine specimens.

NOAA's *Okeanos Explorer* is the first U.S. ship devoted to ocean exploration. Commissioned in 2008, this ship is equipped with remotely operated vehicles to investigate areas of interest and has a sophisticated satellite system to allow communication and interaction with researchers on land.

With a GPS receiver that costs less than a few hundred dollars, you can instantly learn your location—your latitude, longitude, and even altitude—to within a hundred feet or less.

Unmanned submersibles can be operated from the surface ship or from a manned submersible; others are autonomous and can run independent of human interaction for weeks at a time once programmed. These undersea robots go where humans dare not, into tight crevices or sunken ships, and they have the advantage of not needing to eat or sleep.

Satellites transmit data from buoys and other instruments at sea to oceanographers on shore. They also can relay images of the ocean's surface, indicating the location of sea ice, pollution, weather conditions, and, in some cases, ocean currents. For example, scientists may track satellite images of the Gulf Stream over the course of weeks to record changes in the current.

Marine Navigation and GPS

For centuries, navigators and explorers have searched the heavens for a system that would enable them to accurately locate their position on the globe. In 1993, the U.S. Air Force launched the last of 24 satellites into orbit, completing a network known as the Global Positioning System, or GPS.

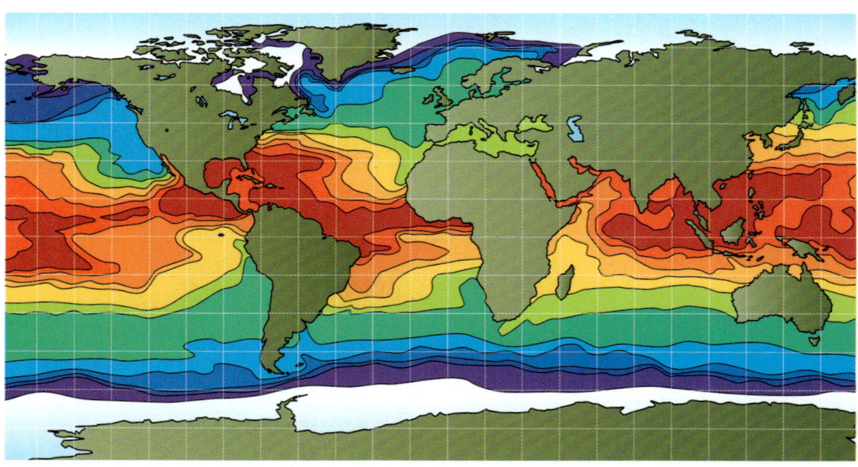

Satellites help oceanographers track changes in such elements as weather patterns, which affect the world's oceans.

This incredible technology was made possible by scientific and engineering advances, particularly the development of the world's most accurate timepieces: atomic clocks, precise to within a billionth of a second. Today, GPS is saving lives, helping society in countless other ways, and generating a multibillion-dollar industry.

Ballard and the *Titanic:* Just the Tip of the Iceberg

As a participant in a 1977 expedition that found hydrothermal vents in the Galapagos Rift, American oceanographer Robert Ballard helped discover the plant and animal life within these deep-sea warm springs at the ocean bottom. To advance deepwater exploration, he designed a series of high-tech vessels, most notably the *Argo-Jason.* This small undersea robot enabled a remote-controlled camera to explore the ocean depths while transmitting live images to scientists aboard a ship. *Argo-Jason* was used to locate the Titanic and numerous other shipwrecks, including the *Lusitania* and *Bismarck.*

At Woods Hole Oceanographic Institute, Ballard was involved in more than 65 expeditions. He helped develop *Alvin,* a submersible equipped with a mechanical arm, used to map the Mid-Atlantic Ridge. He departed to Mystic, Connecticut, in 1997 to lead the Institute for Exploration, a center for deep-sea archaeology.

Careers Related to the Sea

To become an oceanographer, you will need a background in science and mathematics, and a solid knowledge of at least one basic science such as biology, chemistry, geology, or physics. In college, oceanography courses help an undergraduate student learn how science applies to the study of the ocean. Your education should include an undergraduate degree (four years) and at least a master's degree (two more years). You may need a doctorate (three more years).

An oceanographer may choose from several types of careers after completing training. Colleges and universities provide teaching and research opportunities. The government employs oceanographers in areas such as the Department of Commerce and the Environmental Protection Agency. The private sector offers the most opportunities. Engineering businesses, oil and gas extraction companies, and metals mining firms need oceanographers, geologists, and geophysicists.

Oceanographers may detect and track ocean-related weather events, investigate ocean pollution, look for minerals on the seafloor, or study the migration of aquatic animals. Marine engineers design, construct, and repair ships, submarines, and port facilities. Ocean engineers design and install equipment used in the ocean, including oil rigs and other offshore installations, and design breakwater systems to prevent beach erosion.

Plenty of other ocean-related career opportunities exist, including:

- Aquaculture worker
- Coast Guardsman
- Ocean tour or dive operator
- Commercial fisherman or diver, kelp harvester
- Conservationist or fish and game officer
- Lifeguard
- Longshoreman
- Mariculturist
- Meteorologist
- Offshore oil worker
- Sailor, ship's captain, boat operator
- Ship builder
- Underwater photographer

Oceanography Projects

Collecting Plankton

The plankton nets used in most oceanographic research are made of silk or nylon cloth that comes in several different grades and mesh sizes. The holes of this material are very uniform, so it is ideal for accurate collection. However, any fine-meshed cloth such as nylon screening, sheer curtain fabric, cheesecloth, or nylon hose makes a suitable plankton net.

Materials Needed

To make a simple net for requirement 7a, you will need:

- ☐ One leg from an old pair of nylon hose
- ☐ An empty can or plastic bottle (a 16-ounce can will work, or a 2-liter plastic soft drink bottle with the top cut off)
- ☐ Several pieces of medium-weight wire (strong but flexible enough to bend)
- ☐ Large needle
- ☐ Fishing line (nylon thread)
- ☐ Scissors

Create the plankton-collection net by following these steps:

Step 1—Cut the toe out of the nylon hose and push the container into the leg.

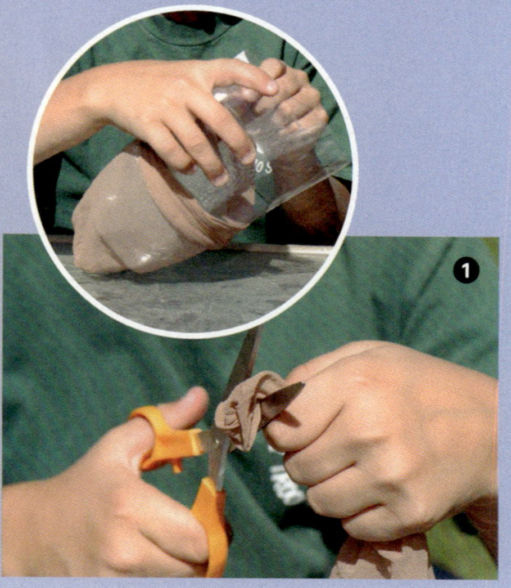

Step 2—Tie a knot in the end of the hose under the collection container.

Another way to attach the collection container is to use a screw-type, adjustable hose clamp. First reinforce the end of the hose by sewing canvas or sailcloth to it.

Then use the clamp to attach the can to the reinforced net. Be sure to use a rigid metal can. The clamp method will not work well with a flexible plastic bottle.

Step 3—Bend a piece of wire into a circle the same size as the top of the hose.

Step 4—Fold the top of the hose over the circle of wire and, using needle and nylon thread, sew the top to the wire ring.

Step 5—Attach the wire ring to the towline by a three-wire bridle. A swivel from a hardware store can be used to help keep the bridle lines from twisting.

Tow the net through the water alongside a dock, wade with it, hold it in a current, or tow it from a rowboat. Continue for about 20 minutes. When you pull the net out of the water, rinse all plankton down the sides of the net into the collection container by pouring water through the outside of the net (not over the top).

After you have concentrated the plankton into the collection container, release the clamp or untie the knot in the hose and remove the container. Study the sample under a microscope or high-power glass. Ask your counselor to help you identify the three most common types of plankton in the sample. A good biology textbook will have pictures that may help you identify your "catch."

Take your measurements at the same spot in the same location every day so you can make accurate comparisons.

Water Testing

For this project, you are to measure water and air temperatures and turbidity four times a day on five consecutive days. As with all scientific experiments, you must use the same exact procedure each time you collect data.

Record the information you gather on an oceanographic log sheet like the one shown here. Make your log sheet large enough to legibly record your observations. Cover your log sheet with a piece of plastic to keep it dry.

Oceanographic Log Sheet

Name:				Location:			
	Water Temperature			Turbidity	Air Temperature	Cloud Cover	Water Roughness
	Surface	Midwater	Bottom				
Date:							
9 A.M.	77° F.	75° F.	71° F.	18 inches	76° F.	partly cloudy	calm
Noon	80° F.	77° F.	73° F.	7 inches	80° F.	partly cloudy	slightly choppy
3 P.M.	81° F.	79° F.	74° F.	8 inches	82° F.	partly cloudy	slightly choppy
6 P.M.	78° F.	78° F.	72° F.	16 inches	75° F.	overcast	calm
Date:							
9 A.M.							
Noon							
3 P.M.							
6 P.M.							
Date:							
9 A.M.							
Noon							
3 P.M.							
6 P.M.							
Date:							
9 A.M.							
Noon							
3 P.M.							
6 P.M.							
Date:							
9 A.M.							
Noon							
3 P.M.							
6 P.M.							

If you do not live near the ocean, you can do this experiment in a lake or stream. You are to measure the water temperature at the surface, in midwater, and at the bottom. However, differences in water temperature and turbidity measurements may not show up in a stream.

First, find out the depth of the water. Then tie a rock to a nylon line and drop it in. When the rock touches the bottom, mark the line with an indelible pen (or pin a clothespin to it) at the surface. Pull up the rock and measure the depth from the rock to the mark. Now you can determine where to take the different water temperature measurements.

Measure the air temperature at the same time you measure water temperature. Note the cloud cover and the roughness of the water.

To measure turbidity (the amount of stirred-up sediment in the water), you will need a Secchi disk, which is used worldwide to monitor the quality of lake water. The disk is 8 inches in diameter, and is painted in black and white quadrants. It can be purchased for about $30 or made from a variety of materials including acrylic, wood, steel, or even an aluminum pie pan or white plastic dinner plate.

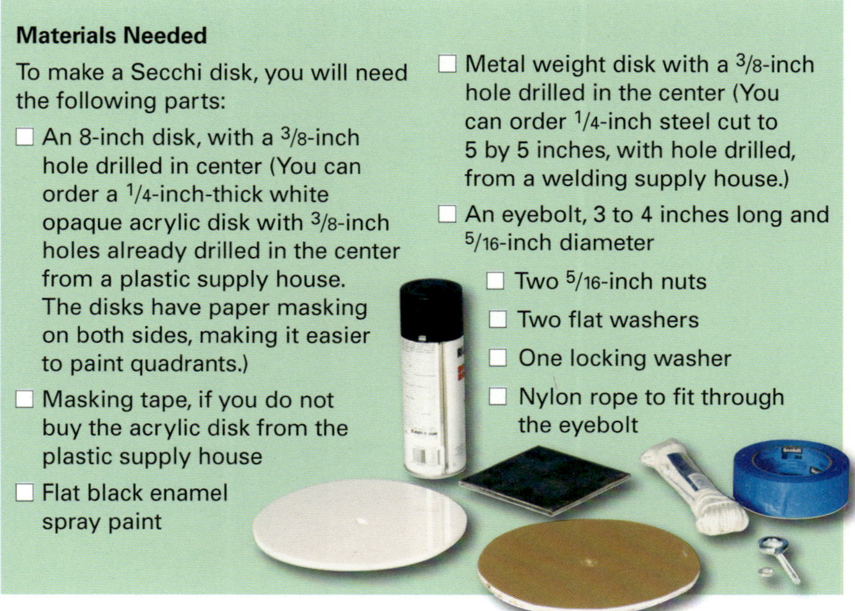

Materials Needed

To make a Secchi disk, you will need the following parts:

- ☐ An 8-inch disk, with a 3/8-inch hole drilled in center (You can order a 1/4-inch-thick white opaque acrylic disk with 3/8-inch holes already drilled in the center from a plastic supply house. The disks have paper masking on both sides, making it easier to paint quadrants.)
- ☐ Masking tape, if you do not buy the acrylic disk from the plastic supply house
- ☐ Flat black enamel spray paint
- ☐ Metal weight disk with a 3/8-inch hole drilled in the center (You can order 1/4-inch steel cut to 5 by 5 inches, with hole drilled, from a welding supply house.)
- ☐ An eyebolt, 3 to 4 inches long and 5/16-inch diameter
- ☐ Two 5/16-inch nuts
- ☐ Two flat washers
- ☐ One locking washer
- ☐ Nylon rope to fit through the eyebolt

Assembling the Secchi Disk

Step 1—Mark the 8-inch disk into four equal quarters. If you are using the acrylic disk with the paper masking, carefully score the lines marked using a craft knife. Then peel off two opposing quadrants. If not using the premasked disk, use masking tape to cover up two opposing quadrants.

Step 2—Spray-paint the exposed quadrants. (Follow the manufacturer's instructions about how to prep the surface and how many coats to apply.) Allow the paint to dry. Peel off the remaining masking tape.

Step 3—Thread a nut and flat washer on the eyebolt, then push the eyebolt through the painted disk and then through the metal weight disk. Thread another flat washer, then the locking washer, and the remaining nut to bolt the metal disk to the painted disk.

Step 4—Tie one end of the nylon rope to the eyebolt.

Secchi Disk Procedure

Step 1—Choose only one method to lower the disk—from a boat or a dock, or by wading. (Anchor the boat to prevent drifting. Be careful not to disturb the water, as that will interfere with the reading.)

Step 2—Do not wear sunglasses. Lower the Secchi disk on the sunny side of the boat or dock. Drop the disk straight down just until it disappears. Mark the rope at the waterline with a clothespin. Slowly pull the rope up until the disk reappears. Mark the rope at the waterline with another clothespin.

Step 3—Determine the midpoint between the clothespins. Measure from that point to the Secchi disk to find how deep you can see into the water. Record that information to the nearest inch on your log sheet.

Several factors affect turbidity: algae, zooplankton, motorboat activity, and soil erosion. What factors are influencing your turbidity readings?

When you have completed your log sheet, show the results on a graph. How does water temperature change with air temperature?

For this project, you are to measure water and air temperatures and compare them. A real oceanographer's log sheet, adapted for this project, is shown earlier in this chapter as a sample for you to use in making your own records.

Measure the water temperature one foot below the surface of a body of water four times daily—at 9 A.M., noon, 3 P.M., and 6 P.M.—for six consecutive days. (You can do this in a lake or stream if you do not live near the ocean, but in most streams, you will not be able to observe variations in water temperature. Can you explain how mixing will keep water temperatures fairly constant near the surface of a stream?)

Measure the air temperature at the same time you measure water temperature. Note the cloud cover, direction and speed of the wind, and the roughness of the water.

Chart your findings on a simple bar graph that shows how the water temperature changes with air temperature.

Wave Testing

A wave generator is a device that is used to produce waves in a tank or other container of water so studies can be made of wave action. For this project, you can make a simple one that will not be expensive or hard to put together. Use a shallow tray or pan for your water container. It should be deep enough to hold at least an inch of water.

Materials Needed

To make the generator, you will need the following parts:

- ☐ One sturdy 9-by-13-inch aluminum baking pan (or something similar)
- ☐ One piece of smooth wood, slightly shorter than the width of the pan (A 1 X 1-inch stock will do; make sure it is smooth by sanding if necessary.)
- ☐ Two brass screw hooks
- ☐ Two rubber bands, at least 4 inches long before stretching
- ☐ One length of string, 4 feet long or so
- ☐ One pinch-type clothespin
- ☐ One ¼-by-1-inch bolt and 2 nuts to fit it
- ☐ One miniature DC motor such as is used to drive model boats (Most hobby and craft shops have these.)
- ☐ One rheostat to control the speed of the motor, available from hobby and craft shops, and flashlight batteries or dry cell to operate the motor
- ☐ A counterweight (such as a bolt or nut) may be necessary

Building the Generator

Step 1—Screw the hooks into the wood about $^{1}/_{2}$ inch from the ends to make a wave bar.

Step 2—Drill a hole in one handle of the clothespin and screw it to the side of the bar, a little off center.

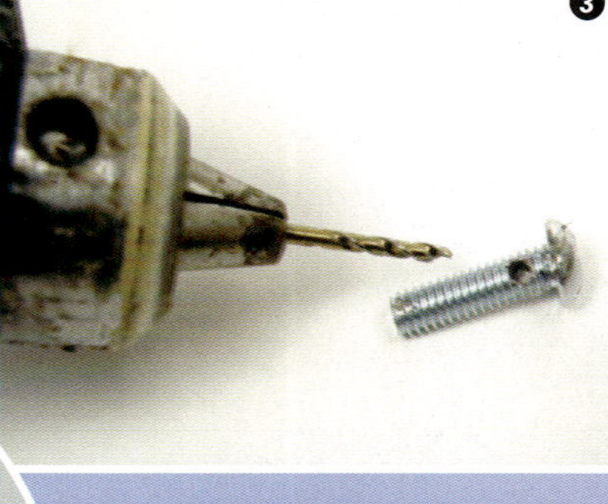

Step 3—Drill a hole in the bolt about $^{1}/_{4}$ inch from the head, just large enough for the motor shaft.

Step 4—Screw one nut on the bolt clear up to the head. Then insert the motor shaft in the small hole and tighten the nut to hold the bolt firmly on the shaft.

Step 5—Put the other nut on the bolt above the shaft, but do not tighten it. It is used to adjust the vibration.

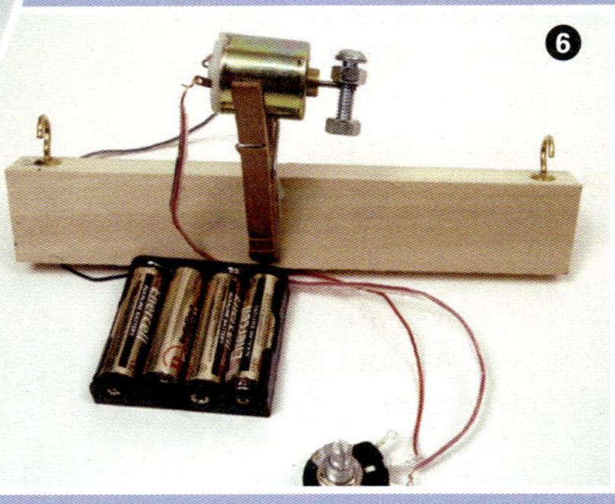

Step 6—Wire the motor to the battery by way of the rheostat and clamp it to the bar with the clothespin. Attach a counterweight if needed.

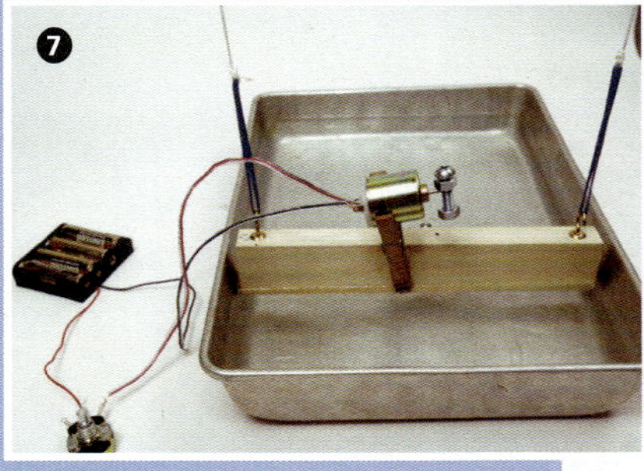

Step 7—Position the generator above the water tray by attaching each end of the string to one of the rubber bands, so the bar just touches the water in the tray and is flat with the water surface. Keep it steady by hooking the string from above to a stable object.

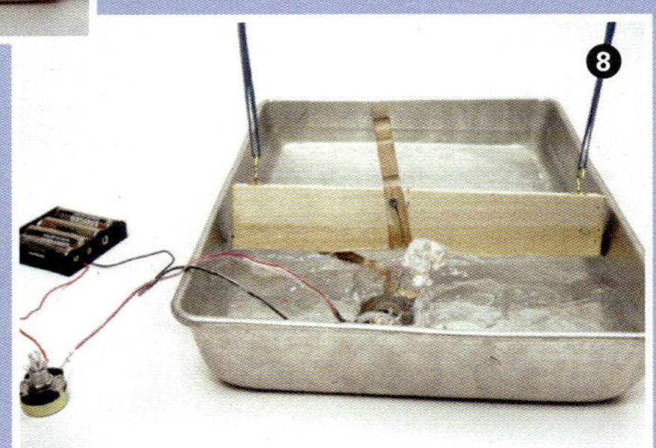

Step 8—Start the generator and adjust the free nut on the 1/4-inch bolt so you can get a vibration of not more than 1/16 of an inch. Too much vibration will spoil the wave patterns.

Adjust the motor speed to achieve the desired wave pattern. If you run the motor too fast, the waves will be too close together; too slow will give the opposite effect.

To see the waves and their effects, shine a bright light on the water so it will be reflected on a wall or large white piece of paper. The best type of light would be a homemade projector or a spotlight, if you have one. By placing small rectangles of glass or plastic in the opposite end of the pan from the generator, you can represent a breakwater, a smooth beach, a jetty, or a groin. Then you can demonstrate how waves are reflected (turned back without bending) or refracted (bent).

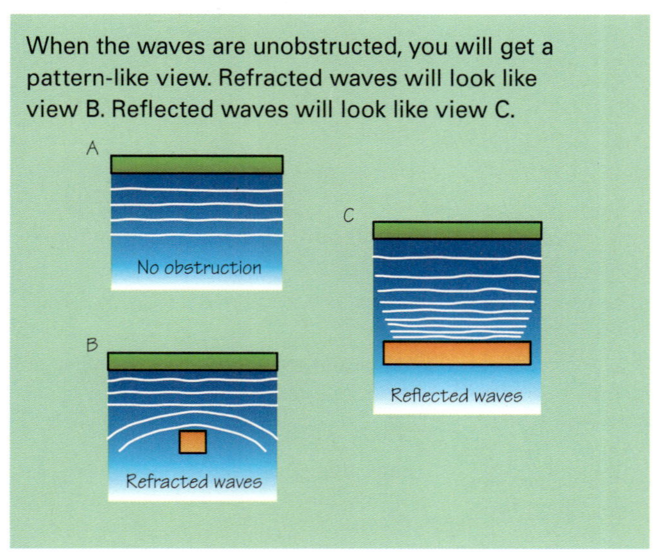

Glossary

abyssal plains. Large flat areas on the ocean floor.

aquaculture. Raising fish or shellfish under controlled conditions for food.

atoll. A ring-shaped coral reef that grows from a submerged volcanic peak.

backwash. Water motion seaward across the foreshore (the portion of the beach from the mean low tide to the top of the ridge above the beach).

baleen. Bony material grown in place of teeth in baleen whales, allowing them to strain food.

barrier beach. A sand ridge extending parallel to a shore, from which it is separated by a lagoon.

barrier island. A broad barrier beach lying parallel to a shore, separated by a sound.

barrier reef. A coral reef that parallels the shore, separated by open water.

beach drift. Movement of material down the beach.

berm. A ridge above the beach formed by storm waves.

breakwater. A barrier that protects a harbor or shore from waves.

chop. Locally generated waves.

cold seeps. Areas where methane- and sulfide-rich fluids seep from the ocean floor.

colonial corals. Coral animals (polyps) that group together to form a coral reef.

continental drift. The slow movement of continents described by plate tectonics.

continental shelf. The part of a continent that extends underwater.

continental slope. The portion of the continental shelf that slopes down to the abyssal plain.

coral. Either the living coral animal (the polyp) or the skeletal remains of the animal.

coral reef. Structure formed when many thousands of coral polyps grow next to and on top of each other in large colonies.

Coral reef

current. A steady movement of water or air in a definite direction.

deep ocean currents. Slow circulation of water at great depths driven by density differences.

density. The average mass per unit volume; a measure of how much matter is squeezed into a given space.

dulse. An edible red alga that grows on rocky shores of the Atlantic Ocean.

foreshore. The portion of the beach from the mean low tide to the top of the ridge above the beach.

greenhouse gas. Gases in Earth's atmosphere that cause the greenhouse effect, or the trapping of heat in the atmosphere.

groin. A structure that angles out into the sea to protect a shoreline.

guyot. Flat-topped seamounts.

holdfast. A structure that attaches certain seaweeds or other algae to the sea bottom or rocks.

holoplankton. Animals that spend their entire life cycle as plankton.

hydrothermal vents. Cracks in the ocean floor from which heated water continually rises.

jetty. A structure such as a pier that projects into a body of water to protect a harbor.

kelp. Various species of large brown algae.

lagoon. A shallow body of water, especially one separated from the sea by sandbars or reefs.

Lagoon

limpets. Mollusks having a cone-shaped shell and sticking to rocks of tidal areas.

lithosphere. Earth's outermost shell.

littoral drift. The combined movement of sediment via longshore drift and beach drift.

longshore current. An offshore current slowing along the shore.

longshore drift. The movement of sediment offshore.

low-tide terrace. The portion of the beach that flattens out where the waves break.

mariculture. Cultivation of marine organisms in their natural habitats, usually for commercial purposes.

meroplankton. Organisms that spend the larval or egg stages as plankton.

mid-oceanic ridges. Underwater mountain ranges generally running north and south in the center of the oceans.

mollusk. Marine invertebrate typically having a protective shell and a soft body.

nodule. Lump of minerals on the ocean bottom.

ocean density. The weight of water divided by the amount of space it occupies.

ocean mixing. Seawater carrying dissolved gases and important nutrients mixing with nutrient-poor seawater.

oceanic fracture zone. Long straight ridges and troughs that run perpendicular to the mid-oceanic ridges.

oceanic trench. A long narrow depression of the seabed with relatively steep sides.

photosynthesis. The process by which green plants, algae, and some other organisms convert the sun's energy into a form of energy that they can use or store.

phytoplankton. Tiny free-floating aquatic plants.

plate tectonics. Earth's outermost shell—the *lithosphere*—is formed of these rigid plates that, over time, slowly move across its surface (as well as under the ocean and under continents) and float on a bed of partly molten rock.

rip current. A current running back out to sea.

salinity. A measure of the amount of dissolved salts in ocean water.

sandbar. Submerged or partly exposed humps of sand or coarse sediment built by waves offshore from a beach. Also called an offshore bar.

sea anemone. Flowerlike marine animal with tentacles surrounding a central mouth.

seafloor spreading. The process of forming new oceanic crust by volcanic material pushing through cracks in the oceanic ridges.

seamount. Large undersea mountain with a submerged peak.

sound. A passage of water between the mainland and an island.

splash zone. In the intertidal zone, the area highest above the waves.

storm surge. Storm-pushed seawater, which may combine with normal tides.

submarine canyon. V-shaped valley cut into the hard rock of the continental slope.

swash. The upward movement of water onto the beach.

swell. Long, far-apart waves in the open ocean.

thermohaline circulation. Vertical movements of ocean water masses caused by density differences that are due to variations in temperature and salinity.

tidal bore. Wave or wall of water that races up an inlet as the tide comes in.

tsunami. A long-period gravity wave generated by a submarine earthquake or volcanic event.

turbidity currents. Currents heavy with sediment that flow down the continental slope.

upper intertidal zone. The tidal zone below the splash zone.

upwelling. A current of cold, nutrient-rich water rising to the surface. Many marine plants and animals live off this water.

vertical migration. The movement at night of some zooplankton to the surface.

westerly. A trade wind that blows surface waters toward the east.

zooplankton. Planktonic animals.

zooxanthellae. A form of algae that lives in corals and other animals and provides food through photosynthesis.

Oceanography Resources

Scouting Resources

Bird Study, Energy, Environmental Science, Fish and Wildlife Management, Fishing, Fly-Fishing, Geology, Mammal Study, Nature, Plant Science, Reptile and Amphibian Study, Soil and Water Conservation, Sustainability, and *Weather* merit badge pamphlets

Visit the Boy Scouts of America's official retail website at http://www.scoutstuff.org for a complete listing of all merit badge pamphlets and other helpful Scouting materials and supplies.

Books

Ballard, Robert D. *Exploring the Titanic.* Econo-Clad Books, 1990.

Broad, William J. *The Universe Below: Discovering the Secrets of the Deep Sea.* Simon and Schuster, 1997.

Carson, Rachel L. *The Sea Around Us.* Oxford University Press, 1991.

Carson, Rachel L., and Sue Hubbell. *The Edge of the Sea.* Houghton Mifflin, 1998.

Center for Marine Conservation Staff. *The Ocean Book: Aquarium and Seaside Activities and Ideas for All Ages.* John Wiley & Sons Inc., 1989.

Coulombe, Deborah. *The Seaside Naturalist: A Guide to Study at the Seashore.* Simon and Schuster, 1992.

Denny, Mark. *How the Ocean Works: An Introduction to Oceanography.* Princeton University Press, 2008.

Earle, Sylvia, and Linda K. Glover. *Ocean: An Illustrated Atlas.* National Geographic, 2008.

Earle, Sylvia A., and Wolcott Henry. *Wild Ocean: America's Parks Under the Sea.* National Geographic Society, 1999.

Ellis, Richard. *Encyclopedia of the Sea.* Alfred A. Knopf, 2000.

Foreman, W. *History of American Deep Submersible Operations.* Best Publishing Co., 1999.

Miller, James W., and Ian G. Koblick. *Living and Working in the Sea.* Best Publishing Co., 1995.

Strang, Craig, Catharine Halversen, and Kimi Hosoume. *On Sandy Shores.* GEMS: Great Explorations in Math and Science, 1996.

Organizations and Websites

American Meteorological Society
Telephone: 617-227-2425
Website: http://www.ametsoc.org

Careers in Oceanography, Ocean Engineering, and Marine Biology
Website: http://www.marinecareers.net

The Discovery Channel Blue Planet
Website: http://dsc.discovery.com/convergence/blueplanet/blueplanet.html

The JASON Project
Telephone: 703-726-4232
Website: http://www.jason.org

National Climatic Data Center
Telephone: 828-271-4800
Website: http://lwf.ncdc.noaa.gov/oa/ncdc.html

National Oceanic and Atmospheric Administration
Telephone: 202-482-6090
Website: http://www.noaa.gov

The Ocean Alliance
Toll-free telephone: 800-969-4253
Website: http://www.oceanalliance.org

The Savage Seas, Public Broadcasting Service
Website: http://www.pbs.org/wnet/savageseas

Scripps Institute of Oceanography
Telephone: 858-534-3624
Website: https://scripps.ucsd.edu

PBS Secrets of the Ocean Realm
Website: http://www.pbs.org/oceanrealm

Acknowledgments

The Boy Scouts of America thanks Thomas Potts, associate director at the National Oceanic and Atmospheric Administration's Undersea Research Center, University of North Carolina Wilmington for his extensive assistance with this revision of the *Oceanography* merit badge pamphlet. The BSA is also grateful to Eagle Scout Lawrence Cahoon, professor, Department of Biology and Marine Biology at the University of North Carolina Wilmington, for his assistance.

The Boy Scouts of America is grateful to the men and women serving on the Merit Badge Maintenance Task Force for the improvements made in updating this pamphlet.

Photo and Illustration Credits

Jupiterimages.com—cover *(diver, fish);* pages 4, 5, 9, 16, 19, 24, 32–35 *(all),* 37 *(top),* 38 *(top),* 40 *(top),* 41 *(bottom),* 42, 45, 48 *(top),* 52 *(bottom),* 59, 61–62 *(all),* 74–80 *(all)*

Life on the Edge 2004 Expedition: NOAA Office of Exploration, courtesy—page 59

NOAA, courtesy—pages 57 *(bottom)* and 58

OAR/National Undersea Research Program (photo by C. Van Dover), College of William & Mary, courtesy—page 53

OAR/National Undersea Research Program; Farleigh Dickinson University, courtesy—page 55

OAR/National Undersea Research Program; University of Connecticut, courtesy—page 56

OAR/National Undersea Research Program; Woods Hole Oceanographic Institute, courtesy—page 54

Pacific Ring of Fire 2004 Expedition: NOAA Office of Exploration; Dr. Bob Embley, NOAA PMEL, Chief Scientist, courtesy—page 60 *(ship workers)*

Shutterstock.com, courtesy—cover (*blue fish*, ©bluehand; *penguin*, ©Chayasit Fangem; *hermit crab*, ©Eric Isselee; *underwater panorama*, ©Isabelle Kuehn; turtle, ©Rich Carey; *two starfish*, ©Paper Street Design; *jellyfish*, ©Jiri Vaclavek); pages 4 (©Richard Whitcombe), 5 (©welcomia), 6 (©Skylines), 9 (©Andrey Yurlov), 16 (©ChrisVanLennepPhoto), 19 (©elgad), 21 (©metrue), 23 (©David Bostock), 24 (©Denis Burdin), 26 (©Pakhnyushchy), 27 (©jo Crebbin), 28 (©panbazil), 29 (Andaman), 30 (©Kritchanut), 32 (©Serg Zastavkin), 34 (©Anna segeren), 35 (©Rich Carey), 37 (*octopus*, ©NaturePhoto; *starfish*, ©Zorandim), 38 (*sea urchin*, ©Lefteris Papaulakis; *crabs*, ©Lau Chun Kit), 40 (*seaweed*, ©Yarlander), 41 (*giant clam*, ©Sean Lema; *Great Barrier Reef*, ©JC Photo), 42 and 43 (©Rich Carey), 45 (©Adwo), 46 (©Thelma Amara Vidales), 48 (©Neirfy), 50 (©Christian Musat), 52 (*jellyfish*, ©saraporn), 61 (©project1photography), 62 (©Daniel Poloha), 74 (©frantisekhojdysz), 75 (*coral reef*, ©Sphinx Wang; *starfish*, ©Fotyma), 76 (*lagoon*, ©crabeg; *octopus*, ©Rich Carey), 77 (©wavebreakmedia), and 80 (©frantisekhojdysz)

Wikipedia.org, courtesy; source: Gulf of Mexico Cruise SJ0107 —page 57 (*Johnson Sea-Link*)

Wikipedia.org/David Rydevik, courtesy —page 18

Wikipedia.org/F.G.O. Stuart, courtesy —page 59 (*Titanic*)

Wikipedia.org/Malene Thyssen —page 48 *(porpoise)*

Wikipedia.org/Whit Welles—page 47

All other photos and illustrations not mentioned above are the property of or are protected by the Boy Scouts of America.

Tim Brox—page 49

Toby Everett—page 46 *(penguins)*

Daniel Giles—pages 15, 39 (*estuary, bed of seaweed*), 60 (*oceanographer on platform; lab worker*), and 70–73 (*all wave generator photos*)

John McDearmon—all illustrations on pages 8, 10, 12–14, 17, 20–22, 40–41, 51–52, 58, and 73